Frontiers of
Scientific Visualization

D1484874

FRONTIERS OF SCIENTIFIC VISUALIZATION

Edited by

Clifford A. Pickover
IBM T. J. Watson Research Center

and

Stuart K. Tewksbury
West Virginia University

A Wiley-Interscience Publication
JOHN WILEY & SONS, INC.
New York • Chichester • Brisbane • Toronto • Singapore

This text is printed on acid-free paper.

Copyright © 1994 by John Wiley & Sons, Inc.

All rights reserved. Published simultaneously in Canada.

Reproduction or translation of any part of this work beyond
that permitted by Section 107 or 108 of the 1976 United
States Copyright Act without the permission of the copyright
owner is unlawful. Requests for permission or further
information should be addressed to the Permissions Department,
John Wiley & Sons, Inc., 605 Third Avenue, New York, NY
10158-0012.

Library of Congress Cataloging in Publication Data:

Frontiers of scientific visualization / edited by Clifford A. Pickover
  and Stuart K. Tewksbury.
      p.   cm.
    "A Wiley-Interscience publication."
    Includes index.
    ISBN 0-471-30972-9 (paper : acid-free paper)
    1. Computer graphics. 2. Visualization. I. Pickover, Clifford A.
II. Tewksbury, Stuart K.
T385.F75 1994
501'.1366—dc20   93-11136
                                                             CIP

Printed in the United States of America

10 9 8 7 6 5 4 3 2 1

Contents

Introduction   1

Clifford A. Pickover

1 Scientific Visualization of Fluid Flow   7

Hassan Aref, Richard D. Charles and T. Todd Elvins

2 Visualization of Scroll Waves   45

Mario Markus and Manfred Krafczyk

3 Visualization of Chemical Gradients   65

Theo Plesser, Wolfgang Kramarczyk and Stefan C. Müller

4 Visualization of Biological Information Encoded in DNA   91

Eugene Hamori

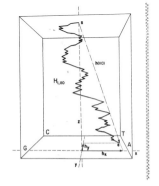

13 OCT

5/11/95

34.95

5519

5 Visualizing Droplet Coalescence Phenomena 123

Paul Meakin, Clifford A. Pickover and Fereydoon Family

6 Computer Simulation of Plant Growth 145

Philippe de Reffye

7 Scientific Display: A Means of Reconciling Artists and Scientists 181

Jean-Francois Colonna

8 Architecture and Applications of the Pixel Machine 213

Michael Potmesil and Eric M. Hoffert

9 Brave New Virtual Worlds 245

David M. Weimer

HELMET-MOUNTED HEAD AND EYE TRACKER

FOVEA PROJECTOR

BACKGROUND PROJECTOR

Index 279

Frontiers of
Scientific Visualization

Introduction

Clifford A. Pickover
IBM Watson Research Center
Yorktown Heights, NY

"The standard argument to promote scientific visualization is that today's researchers must consume ever higher volumes of numbers that gush, as if from a fire hose, out of supercomputer simulations or high-powered scientific instruments. If researchers try to read the data, usually presented as vast numeric matrices, they will take in the information at a snail's pace. If the information is rendered graphically, however, they can assimilate it at a much faster rate."
  R. Friedhoff and T. Kiley, *The eye of the beholder*, 1990.

"Visualization is a method of computing. It transforms the symbolic into the geometric, enabling researchers to observe their simulations and computations. Visualization offers a method for seeing the unseen. It enriches the process of scientific discovery and fosters profound and unexpected insights. In many fields it is already revolutionizing the way scientists do science."
  B. McCormick, T. DeFanti, and M. Brown, 1987.

"We take a handful of sand from the endless landscape of awareness around us and call that handful of sand the world."
  Robert Pirsig.

In October 1986, the National Science Foundation sponsored a meeting of a *Panel on Graphics, Image Processing, and Workstations* to help establish priorities for acquiring graphics hardware and software

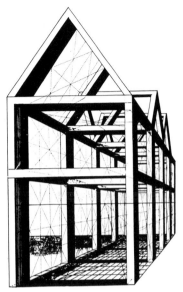

Figure A

at research institutions doing advanced scientific computing. The application of graphics and imaging techniques to computational science was a new area of endeavor which Panel members termed *Visualization in Scientific Computing* (ViSC). The Panel maintained that scientific visualization was emerging as a major computer-based technology requiring significantly enhanced federal support. Existing tools were simply not adequate to meet the needs of supercomputer centers around the country.

The first workshop on ViSC in 1987 brought together researchers from academia, industry, and government. Their report made it clear that visualization has the potential for fostering important scientific breakthroughs. The significance of scientific visualization has grown since then. As Richard Hamming observed many years ago, "The purpose of [scientific] computing is insight, not numbers.

Computer graphics has become indispensable in countless areas of human activity – from colorful and lighthearted television commercials, to strange new artworks, to evolutionary biology, to processed images from the edges of the known universe. As additional background, the term *scientific visualization* has come to mean the marriage of high-speed computation and colorful 3-D graphics to help researchers better understand complicated models. *Scientific visualization is the art of making the unseen visible.* Aside from the applications above, scientific visualization is also becoming very useful to mathematicians. However, long before computer graphics, pictures and physical models played an important role in mathematics. Nineteenth-century mathematicians, for instance, regularly drew pictures (Figure A) and sculpted bizarre plaster or wooden models to help them visualize and understand geometric forms.

We have come a long way from these pencil-and-paper mathematical diagrams. How will graphics supercomputers of the future affect mathematics and science? As just one example, today mathematicians use computer graphics to add a vivid new dimension to geometric investigations (Figure B). One center of this kind of activity is the Geometry Supercomputer Project, based at the University of Minnesota at Minneapolis-St. Paul. Established in 1987 by a group of thirteen mathematicians and computer scientists, this project addresses various unsolved mathematical problems using powerful computers. As you read through the papers in this book, you are urged to think towards the future and consider how the applications in this book will be affected by faster and faster graphics.

So extensive is the scientific world's interest in visualization, that

Figure B
Scientific visualization can reveal the beautiful and intricate behavior of mathematical recursion. I produced this *Golden Julia Set* by iterating $z = z^2 + i/\phi$ where ϕ is the so-called golden ratio of the ancient Greeks ($\phi = 1.68034$). The interior concentric rings are contours corresponding to the minimum value of an orbit as an initial point is iterated. (From *Computers and the Imagination* by C. Pickover. ©1991 by St Martin's Press. All rights reserved.)

keeping up with the literature on the subject is rapidly becoming a full-time task. In 1989 the world's scientific journals published about 300 articles with the words "visualization" or "visualizations" in the title. Figure C shows the number of papers with titles containing these words for the years 1975-1990, the 1990 value estimated from data for January-June 1990.

A recent National Science Foundation study found that the sciences were in urgent need of government support for graphic tools to view the millions of bytes of data that computers are heaping upon researchers (Wolff, 1988). The commercial world is beginning to recognize the visualization needs of the scientific community and respond to them. Graphics workstations were born more than five years ago to give investigators more comprehensible representations of their results. Manufacturers such as Silicon Graphics, IBM, PIXAR, and many others have aimed to bridge the gap between fast computation and 3-D images.

In the chapters which follow, researchers have used both simple 2-D graphics and more sophisticated 3-D graphics to help visualize a range of phenomena. The usefulness of a particular graphic repre-

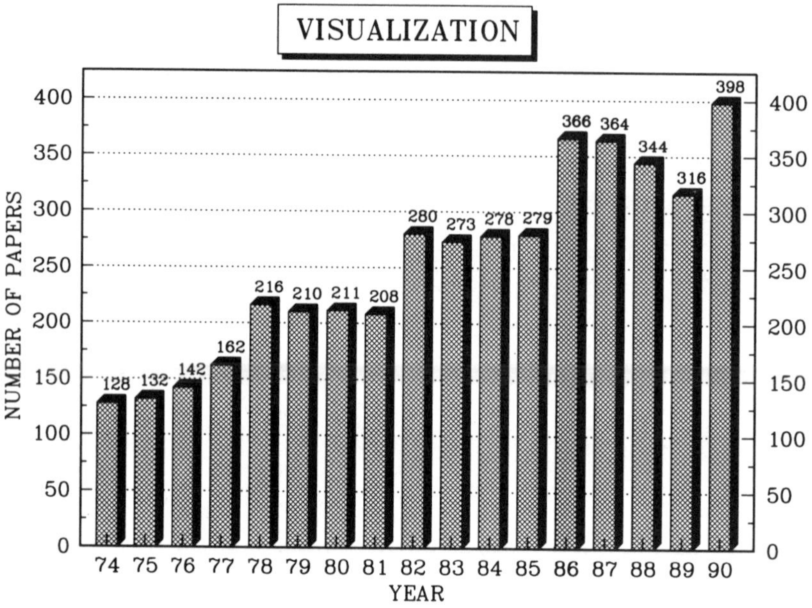

VISUALIZATION

Figure C
A review of the world scientific literature shows the number of visualization articles rising during the years 1975 - 1990. (From *Computers and the Imagination* by C. Pickover. ©1991 by St Martin's Press. All rights reserved.)

sentation is determined by its descriptive capacity, potential for comparison, aid in focussing attention, and versatility. For background articles in the field of scientific visualization, see the reference section which follows. There are also many journals devoted to the subject of scientific visualization, including: *Computers and Graphics* (Pergamon), *The Visual Computer* (Springer-Verlag), *IEEE Computer Graphics and Applications* (IEEE Computer Society), *The Journal of Visualization and Computer Animation* (Wiley), *Computer Graphics World* (PennWell Publishing), and *Pixel* (American Association of Computing Machinery). References 12-15 are four favorites in terms of visualization applied to unusual and specific areas of science and art.

REFERENCES

[1] R. Friedhoff and T. Kiley, The eye of the beholder, *Computer Graphics World*, **13**(8), 47-56 (1990).

[2] I. Peterson, The color of geometry, *Science News*, **136**, 406-410 (Dec. 23, 1989).

[3] E. Tufte, *The Visual Display of Quantitative Information*, Graphics Press, Connecticut (1983).

[4] R. Wolff, The visualization challenge in the physical sciences, *Computers in Science,* **2**(1), 16-31 (1988).

[5] C. Pickover, *Computers, Patterns, Chaos, and Beauty,* St Martin's Press, New York (1990).

[6] C. Pickover, *Computers and the Imagination,* St Martin's Press, New York (1991).

[7] H. Wainer and D. Thissen, Graphical data analysis, *Annual Review of Psychology,* **32**, 191-241 (1981).

[8] B. Mandelbrot, *The Fractal Geometry of Nature,* Freeman, New York (1982).

[9] R. Friedhoff and W. Benzon, *Visualization: The Second Computer Revolution,* Abrams, New York (1989).

[10] M. Lynch and S. Woolgar, *Representation in Scientific Practice,* The MIT Press, Cambridge, MA (1990).

[11] E. J. Wegman, Hyperdimensional data analysis using parallel coordinates, *Journal of the American Statistical Asociation,* **85**, 664-675 (1990).

[12] P. Gerdes, Reconstruction and extension of lost symmetries: examples from the Tamil of South India, *Computers and Mathematics with Applications,* **17**, 791-813 (1989).

[13] H. Bohr and S. Brunak, Complex Systems, **3** 9 (1989).

[14] P. Desain and H. Honig, Quantization of musical time: a connectionist approach, *Computer Music Journal,* **13**, 56-66 (1989).

[15] J. Rangel-Mondragon and S. J. Abas, Computer generation of penrose tilings, *Computer Graphics Forum,* **7**, 29-37 (1988).

[16] B. McCormick, T. DeFanti, and M. Brown, Visualization in Scientific Computing, *Computer Graphics (AC-SIGGRAPH),* **21**(6), 1-15 (1987)

1

Scientific Visualization of Fluid Flow

Hassan Aref, Richard D. Charles and T. Todd Elvins
San Diego Supercomputer Center
University of California at San Diego
La Jolla, California

1.1 INTRODUCTION

Scientific visualization is a new, exciting field of computational science spurred on in large measure by the rapid growth in computer technology, particularly in graphics workstation hardware and computer graphics software. Pictures of breathtaking realism and photographic detail can now be generated in minutes or hours on computer displays that are often within financial reach of individuals and certainly within reach of small companies and academic institutions. These tools are beginning to impact our daily lives through usage in the arts, particularly film animation, and they hold great promise for scientific research and education. When computer graphics is applied to scientific data for purposes of gaining insight, testing hypotheses, and general elucidation, we speak of *scientific visualization*.

As a field of science, visualization is still very much in its infancy and many have probably wondered whether considering it as a new field of science is even legitimate. One may point here to an interesting parallel that occurred several centuries ago when artists, Albrecht Dürer (1471-1528) and his school in particular, began to inquire into the laws of depth perception and perspective. As is well known, in the hands of Desargues, Poncelet, Pascal, Brianchon and others, this inquiry led ultimately to the formulation of the beautiful subject in mathematics that we today know as *projective geometry*. It is not unreasonable to expect that similar comprehensive developments will arise from the practice of scientific visualization and the large number of algorithms and mathematical results to which this activity gives rise.

In the arena of fluid flows, the practice of visualization, some of it originally motivated by science but much of it motivated by artistic desires, may be traced at least as far back as Leonardo da Vinci (1452-1519). While artists have always been attracted by naturally occurring fluid flows that – so to speak – provided their own visualization, such as breaking waves, rapidly running streams, and atmospheric clouds, da Vinci apparently was one of the first to use visualizing agents, in the form of fine particles of sand or dirt and wood shavings, to seed a flow, and then to sketch the patterns that he saw (Figure 1.1). In many ways, what he did, using very primitive means and complemented by his outstanding artistry, captures the activity of flow visualization as it is practiced today. Certainly the particles that are placed in a flow to be visualized are now much more refined[1] and the recording methods are technologically advanced, typically involving some form of strong illumination, such as a sheet of laser light, combined with still or film/video photographic recording.

High-speed photography as exemplified in the work of Edgerton[2] has allowed us to capture flow features that even the most adept visual observer would probably miss (see reference [1] for a collection of photographs several of which involve fluid flows). The phenomenon of splashes was thoroughly investigated using high-speed photography by Worthington [2] in a now classical study. Another seminal example using high-speed photographic techniques is the 1886 visualization by Mach of the shock wave due to a bullet traveling in air [3].

The injection of the computer into the process of converting "data" to "image" in the analysis of laboratory experiments provides a highly adaptable "lens" to the experimenter's camera. On the other hand, in the numerical simulations that we encounter in computational fluid dynamics (CFD), the "data" are simply a long list of numbers, and the

Figure 1.1
Sketches from Leonardo da Vinci's *Del moto e misura dell'acqua* of eddies and wave patterns from various flow situations.

---

[1] Smoke, dye, small gas bubbles, aluminum flakes, etc., are usually introduced with great attention to uniformity and continuity.

[2] His picture of a splash by a drop in a shallow pan of milk is famous.

conversion of this to an image gives to the computer a unique role in the visualization process, since at the level of an image no "objective reality" exists before the image is created by a computer program. If we were to compare the creation of a flow visualization image in a laboratory experiment to the recording of a musical performance (which, incidentally, also involves considerable intervention of electronics), then the computer generated image created from numerical simulation data would have to be compared to an actual performance that brings forth sound on the basis of a sheet with musical notation.

Whereas scientific visualization in general may be a very young science, flow visualization (and we are here thinking primarily of the visualization of laboratory flows) is a relatively mature subject. The interested reader can find chapters on it in numerous general texts devoted to fluid mechanics (see, for example, the discussion by Tritton [4]). Specialized treatments of the subject also exist (e.g., Werlé [5], Asanuma [6] and Merzkirch [7]). The film by Kline [8] on the subject should also be noted. Particular mention must be made of the collection of black and white photographs edited by Van Dyke [9] summarizing a century of laboratory flow visualization studies. This little volume has appealed to scientists and others working well outside the usual bounds of fluid mechanics. The collection published by Van Dyke is supplemented annually by publication in the journal *Physics of Fluids* of the winning entries in the "Gallery of Fluid Motion" photo contest (see Reed [10-15]) held in conjunction with the annual meeting of the Division of Fluid Dynamics of the American Physical Society (the major recurring conference event in the field of fluid mechanics in the United States). Again, essentially all the pictures displayed are from laboratory visualizations, although several of these involve considerable computer processing.

Questions of technique and laboratory implementation aside, the issue in flow visualization is basically that of elucidating the geometry and topology of certain fields. The most immediately available field in a fluid is the velocity field, and the problem of flow visualization thus necessitates a careful look at flow kinematics. This, in itself, is a most interesting subject which, quite possibly, can play for flow visualization precisely the role that projective geometry played for the study of perspective. We devote Section 1.2 to a summary of the pertinent issues.

Because of the long track record in visualizing laboratory flows, there is considerable "empirical material" on which to draw when discussing the subject. In Section 1.3, we have collected a sequence of pictures noteworthy not only for the message about fluid dynamics that they impart, but also for the range of techniques that they represent. In assembling this "picture gallery," we have freely intermingled laboratory and CFD images and we have purposefully juxtaposed images of very different "styles."

Once the tour in Section 1.3 is completed, we embark in Section 1.4 on a discussion of some general issues: the challenges faced by different dimensionalities and tensorial ranks of the quantities to be visualized and the related issue of the use of color. We comment on the issues of resolution and of visualizing complex, three-dimensional structures. Our conclusions are collected in Section 1.5.

1.2 PARTICLES AND FIELDS

A typical task of flow visualization is to elucidate the velocity field, which we represent by the symbol $V(x, t)$, in the fluid. This is a three-dimensional vector field dependent on space (x) and time (t). The available tools include (a) direct point measurements of the velocity and (b) seeding the flow with particles of some kind that are then tracked and photographed. The latter procedure usually gives more global data because several particles can be released at once. However, one reason that such flow visualizations are difficult to interpret is that the relationship of individual particle motions to the underlying flow field can be very complicated. To understand this, let us first review what is meant by the velocity field in a fluid and how it determines particle motions within it.

1.2.1 A Primer on Flow Kinematics

In the mechanics of a finite collection of particles, we are used to the notion that the position of each particle, labeled for convenience by a number $\alpha = 1, \dots, N$, is given by two or three coordinates collected in a position vector $X_\alpha(t)$ that depends on time. To describe how a configuration of particles changes with time, we introduce the particle velocities $u_\alpha(t) = \dot{X}_\alpha(t)$, where the overdot signifies a time derivative. When we come to solve a problem in particle mechanics, the initial condition is given in terms of the values at time $t = 0$ of all the position vectors and all the particle velocities. Newton's laws then allow us to update the particle configuration using the forces acting on the particles to determine their accelerations.

For the mechanics of a fluid, some generalization of this description is required. First of all, there is now a continuum of particles rather than a finite number. The discrete, integer-valued label must therefore be replaced by a continuously variable label, which is often conveniently taken to be the position of each point in the fluid continuum at the starting time $t = 0$. We can then consider the coordinates of all the fluid particles, $X(t; a)$, where the label a is the initial position, that is,

$$a = X(0; a)$$

The particle velocity is similarly given by

$$u(t; a) = \left(\frac{\partial X}{\partial t}\right)_a \tag{1.1}$$

The subscript on the partial derivative reminds us that it is to be evaluated following this particular particle with label a.

It turns out that the particle description implied by (1.1) is often not the best to consider when writing down dynamical equations for the fluid. It is more useful, as first shown by Euler, to consider the changing velocity at a fixed point in space, even though this velocity will be due to different fluid particles at different times. The aggregate of these velocity vectors at different spatial points and different times makes up the velocity field $V(x, t)$. The advantage of this concept is that the field may be steady, that is, V will not depend explicitly on time, even though a considerable amount of motion is going on. For example, fluid in uniform rotation has a steady velocity field even though every particle of the fluid is, of course, undergoing uniform circular motion. History has attached the names "Eulerian" to field quantities and "Lagrangian" to quantities pertaining to individual particles.

Formally, the connection between the velocity field, $V(x, t)$, and the particle velocities, $u(t; a)$, is that the field value at a certain point in space and a certain instant in time reflects the velocity of the particle that happens to be passing through that space-time point. In particular,

$$u(t; a) = V(X(t; a), t) \tag{1.2}$$

Combining (1.1) and (1.2) we obtain

$$\left(\frac{\partial X}{\partial t}\right)_a = V(X(t; a), t) \tag{1.3}$$

or, reducing the label a to the status of an initial condition for this system of ordinary differential equations, simply

$$\frac{dx}{dt} = V(x, t) \tag{1.4}$$

We shall refer to (1.4) as the *advection equations*. Sufficiently light and inert particles move according to (1.4) once the flow field **V** is given.

1.2.2 Pathlines, Streaklines, Timelines and Streamlines

The problem from the point of view of flow visualization is that solutions to the advection equations (1.4) do not, in general, directly give the desired flow field. We shall ignore here the issue of identifying particles for which (1.4) is the equation of motion, and consider this an experimental challenge. The mathematical problem is well-defined, and in CFD there is, of course, no such problem. It is usually easy for an experimenter to seed a flow with particles and obtain analog solutions of (1.4). It is much more difficult from this information to deduce properties of the changing field **V**. In CFD the problem is typically different. In most computational investigations, the objective is to calculate the field **V** (and maybe other fields such as pressure, temperature, density, etc.). Hence, the output of a CFD investigation will usually include some approximate form of $V(x, t)$. Predicting particle tracks is then a separate calculation. The problems arise in comparing experimental and numerical data, where, crudely speaking, the experimenter can easily achieve voluminous data on advected particles, but only much more tediously can he achieve extensive data on field values.

Assume that a particle is released in a flow at time $t = t_0$ and position $x = x_0$. The ensuing trajectory of the particle is called a *pathline*. This is really just the graph of the function $X(t; a)$ introduced before, that is, the pathline is the solution of (1.4) subject to an initial condition of the form: $x = x_0$ for $t = t_0$. Segments of pathlines are used frequently for visualization. If the fluid has a free surface, a few buoyant, light-reflecting particles can be sprinkled onto it and their motion tracked photographically. Many older flow visualizations employ this simple technique. Figure 1.2a shows an example elucidating the famous von Kármán vortex street formed behind a cylinder (Werlé [5]). In most pictures of the vortex street, the camera is stationary and the cylinder moves by it. In this picture, the camera has been translated such that the vortex wake pattern appears stationary. Figure 1.2b shows a comparison plot generated from the original von Kármán point vortex street model. Considerable agreement in the flow patterns is achieved. (The color representation in Figure 1.3 is discussed later.)

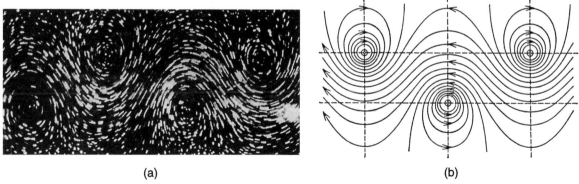

(a)   (b)

Figure 1.2
Two visualizations of the von Kármán vortex street. (a) Light reflecting particles have been sprinkled on the free surface and photographed in a finite-time
exposure producing pathline segments. (b) The actual pathlines. See insert
for color representation.

Figure 1.3
A representation of the flow in
which the color gives the
direction of the velocity vector
and the intensity gives its
magnitude. See insert for color
representation.

Next, assume that instead of releasing one particle at some position we continuously release particles at that position by a "bleed" arrangement. The different particles have trajectories given by solving (1.4) with initial conditions $\mathbf{x} = \mathbf{x}_0$ for $t = t_r$, where the release times, t_r, span some continuous interval. The simultaneous positions of all these particles make up what we call a *streakline*. Streaklines are also used frequently for laboratory flow visualization. Models of airplanes, missiles, cars, trucks, ships and buildings are outfitted with holes through which colored smoke can be released and are then placed in a wind or water tunnel for testing. The flow pattern is obtained by recording the streaklines. See Werlé [5] for examples of this type of visualization.

A *timeline* is defined as follows: Release a "batch" of particles at some instant. The position of these particles at any later instant defines the timeline. The particles typically are arranged initially in a line or in a blob of some particular shape (e.g., a sphere). The timeline may be considered to be the "dual" of the streakline in this sense: The timeline is made up of particles started at the same time at different positions; the streakline, on the other hand, is made up of particles started at the same position at different times. Figure 1.4, from Corrsin and Karweit [16], shows an example of timelines in a turbulent flow. The flow is from left to right. Timelines of hydrogen bubbles (actually double lines) are generated by passing a periodic electric pulse through a fine platinum wire in a water tunnel. The wrinkling of the lines is due to the turbulent flow.

As a final type of "line," we define the streamline. This is simply a field line (in the sense of Faraday) for the velocity field - that is, a curve with the property that the velocity field is everywhere tangent to it. If an infinitesimal displacement along the tangent to the streamline is $d\mathbf{x} = (dx, dy, dz)$, the statement that this vector is parallel to $\mathbf{V}$ gives the equation for the streamline in the form

$$\frac{dx}{u} = \frac{dy}{v} = \frac{dz}{w} \tag{1.5}$$

where $\mathbf{V} = (u, v, w)$. Note that streamlines are only defined using instantaneous values of the field. For a time-dependent field, the entire pattern of streamlines varies from moment to moment (and the concept is, in fact, not particularly useful).

A basic result of the subject is the following:

Theorem: *For steady flow streaklines, pathlines and streamlines coincide.*

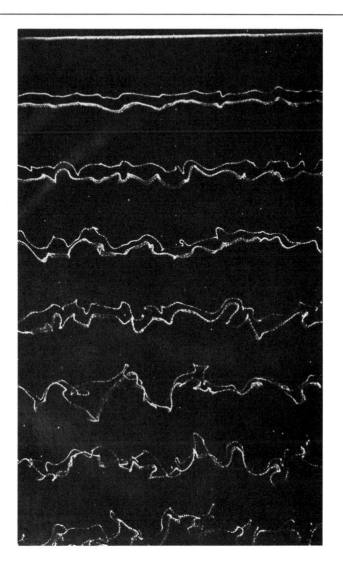

Figure 1.4
Timelines of hydrogen bubbles
in a turbulent flow (from
Corrsin and Karweit [16]).

Proof: Consider the pathline through a given point. Clearly the particle will move along a streamline initially and thus, if the velocity field is not changing in time, it will continue to move along the same streamline. This shows that pathlines are streamlines. Similarly, if the field is steady (i.e., time-independent), a particle released moments after another as part of a streakline will proceed on exactly the same path, which is, in fact, the pathline of the earlier particle. This shows that streaklines are pathlines.

Hence, for the case of steady flow, the release of particles into the flow, whether to produce pathlines or streaklines, will end up delineating the desired streamlines. Indeed, in Figure 1.2b the curves plotted are the streamlines whereas the segments in Figure 1.2a are pieces of pathlines.

1.2.3   Chaotic Advection

It is essential to make some qualifying remarks concerning the theorem just stated. First, it is important to note that, even though the various path/streak/streamlines coincide in a steady flow, there is no implication that they will be geometrically "simple." In fact, there is evidence to suggest that, even in three-dimensional steady flows, the path/streak/streamlines can be space-filling curves!

Second, the harmonious state of affairs that exists for steady flows is completely false for unsteady flows. For a general time-dependent flow (even in two dimensions), pathlines are different from streaklines and both are different from streamlines. These differences have little to do with the amount of agitation of the fluid. For example, it is possible to have substantial deviations of these three types of lines in what is known as *Stokes flow*, where the effects of fluid inertia are insignificant compared to pressure and viscous forces and where the Reynolds number (a key measure of the "nonlinearity" of a fluid motion) is tiny.

The reason for both of these remarks is that the advection equations (1.4) are sufficiently rich that they may display chaotic solutions for particle motion even though the velocity field, which appears on the right-hand side, is very "simple" [17]. To cite an example pertaining to the first remark, consider the so-called *ABC flow*

$$\mathbf{V}(\mathbf{x}) = (A \sin z + C \cos y, B \sin x + A \cos z, C \sin y + B \cos x) \qquad (1.6)$$

defined on a cube of side 2π. This field satisfies the equations of motion for steady, inviscid, three-dimensional flow (i.e., the three-dimensional Euler equations) for any values of the constants A, B, and C. Figure 1.5 illustrates a path/streak/streamline of this flow [18,19]. In the panels labeled B, C, and D, three views are given of this curve as it appears if seen through one of three perpendicular faces of the basic periodic cube containing the flow. Panel A showns a cut, parallel to one of the cube faces, through the curve. It is clear that a very complicated curve is being produced, which as time goes on appears to fill a finite volume within the cube. The appearance of

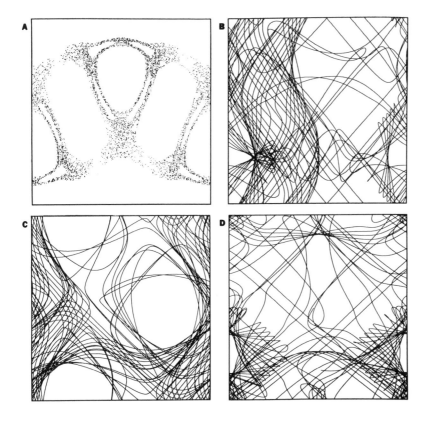

Figure 1.5
"Poincaré section" (A) and
pathlines (or streamlines)
(B,C,D) in chaotic ABC flow.

chaos in this model flow was first suggested on the basis of analysis and numerical experiments by Arnol'd [20] and Hénon [21]. A recent detailed study of it may be found in reference [22].

Similar results pertain to the case of time-dependent flow where even in two dimensions particle tracks can be completely erratic. Figure 1.6 illustrates this with a visualization of a different kind [18,19]. Imagine that we have a large shallow pool of fluid and that we are considering draining it through two sinkholes placed some distance apart. For added realism, assume that, whenever one of the sinks is opened, a swirling motion is set up with a vortex of a certain strength centered on the sinkhole. Let the strength of the sink (in two dimensions measured in area drained per time unit) be Q, and let the strength of the vortex be Γ (the appropriate measure is the *circulation*, which has the same units as Q of length squared over time). If both sinks are left open all the time, a simple flow pattern results in which the boundary between fluid exiting one sink or the other is

Figure 1.6
Color-coded drainage basins
and exit times for the
two-sink/vortex model. See
insert for color representation.

a smooth curve. However, if the sinks are opened and closed alternately, the motion of particles in the drainage flow becomes chaotic and extremely complicated patterns of motion arise. The system is determined up to scale by two dimensionless parameters. One is simply the ratio $\alpha = \Gamma/Q$; the other is a measure of how much fluid the sink takes in during one opening: If the sink-to-sink distance is a, and the time either sink is open is T, then $\lambda = QT/a^2$ provides the required nondimensional measure.

In Figure 1.6, which corresponds to one setting of α and λ, each pixel of the square shown was queried in order to determine when the fluid particle at its location would exit and through which sink. For particles exiting through the sinkhole on the left (the first to be opened), colors in the reds and yellows were used to differentiate exit times. For particles exiting through the sinkhole on the right, blue colors were used. The interleaved striations in this picture, the pattern of which changes dramatically as the governing parameters α and λ are varied, indicates a highly sensitive dependence on initial position resulting from the chaotic nature of the motion in this flow. This type of visualization, which would be rather difficult to do in the laboratory, carries important messages with regard to transport. We shall call it *point of origin coding*. The figure also shows that many of the surfaces to be visualized in fluid flow are highly complex, and in general probably fractal.

1.2.4 Vortex Dynamics

As a final topic, closely related to the above, we mention the dynamics of the vorticity field, formally defined as the curl of the velocity field ($\nabla \times \mathbf{V}$) and usually designated by the symbol $\vec{\omega}$. The interpretation of the vorticity is that it gives the magnitude and direction of the angular velocity vector for each particle of the fluid (actually, the vorticity magnitude is *twice* the angular velocity). Classical fluid mechanics operated, in essence, in a world without vorticity. Nevertheless, the laws of how vorticity evolves during fluid motion were established more than 125 years ago in a seminal paper by Helmholtz [23]. The theory of vorticity dynamics is of profound importance to modern flow visualization because it turns out that the vorticity field is often very accessible to direct visualization.

The vorticity equation, as it is sometimes called, follows by taking the curl of the Navier-Stokes equation (the momentum equation for the fluid, i.e., essentially the analog of Newton's second law) and, for an incompressible fluid, reads

$$\frac{d\vec{\omega}}{dt} = \vec{\omega} \cdot \nabla \mathbf{V} + \nu \nabla^2 \vec{\omega} \tag{1.7}$$

where the derivative on the left-hand side is evaluated following the motion of the fluid particle, $\mathbf{V}$ denotes the velocity field, and ν is the so-called kinematic viscosity (the ordinary viscosity divided by the fluid density). The second term on the right-hand side of (1.7) is a diffusive term, which lends the interesting physical interpretation to the kinematic viscosity of a diffusion constant for vorticity. Broadly speaking, the action of this term is easy to understand.

The interesting feature that Helmholtz noticed of the (inviscid flow) equation made up of the balance between the time derivative on the left and the first term on the right of (1.7) is that it has precisely the same form as the advection equation for a small line element! To see this, consider tracking two particles using (1.4), one at $\mathbf{x}$, the other a small distance away from x, say at x+δs. Consider a short time interval Δt. The particle at x moves to $\mathbf{x} + \mathbf{V}(\mathbf{x}, t)\Delta t$. Similarly, the particle at $\mathbf{x} + \delta\mathbf{s}$ moves to $\mathbf{x} + \delta\mathbf{s} + \mathbf{V}(\mathbf{x} + \delta\mathbf{s}, t)\Delta t$, which may be expanded to first order in the small quantity δs as

$$\mathbf{x} + \delta\mathbf{s} + \mathbf{V}(\mathbf{x}, t)\Delta t + \delta\mathbf{s} \cdot \nabla\mathbf{V}(\mathbf{x}, t)\Delta t$$

Hence, the vector from one particle to the other, which originally was just δs, has a short time Δt later changed to $\delta\mathbf{s} + \delta\mathbf{s} \cdot \nabla\mathbf{V}(\mathbf{x}, t)\Delta t$. In other words the rate of change of δs following the fluid motion is $\delta\mathbf{s} \cdot \nabla\mathbf{V}(\mathbf{x}, t)$, or

$$\frac{d(\delta\mathbf{s})}{dt} = \delta\mathbf{s} \cdot \nabla\mathbf{V} \tag{1.8}$$

The correspondence between (1.7) (with $\nu = 0$) and (1.8) is usually stated as "vortex lines are material lines," or in the words of Tait's translation of Helmholtz's 1858 paper [23]: "Each vortex-line remains continually composed of the same elements of fluid, and swims forward with them in the fluid." Thus, the rising column of visible debris actually marks a tornado or waterspout, and the vortex is where this "visualization" suggests it to be even as it moves!

Clearly, this ability to "tag" vortices is very important for flow visualization. Vortices are usually dynamically significant features of a flow and knowing where they are and, how they are moving can answer many of the questions that provided the objective for conducting the flow visualization in the first place. Furthermore, providing a point measurement of the vorticity field is a complicated task that can only be accomplished approximately after considerable labor.

1.3 A PICTURE GALLERY OF VISUALIZATIONS

We now show a sequence of flow visualization images, commenting briefly on content and technique. Clearly such a sequence must be both biased and incomplete in terms of coverage. The objective is not to make value judgments but to impart to the reader both (a) a sense of the range of flow visualization activities that go on in current research and (b) an appreciation for the types of questions being asked.

1.3.1 Stokes Flow over a Cavity

In the limit of very slow flow, or equivalently flow of a very viscous fluid, the equations of motion simplify and become linear partial differential equations for the velocity field. This flow regime is known variously as "creeping flow," "linearized viscous flow," or "Stokes flow." Nevertheless, many complicated flow features arise in this technologically important regime, the elucidation of which has occupied researchers for years. An interesting example is shown in Figure 1.7. We are concerned with steady Stokes flow over a groove or cavity in a flat plate and, specifically, with the issue of the flow pattern within the groove. Laboratory visualization photographs by Taneda [24] show particle path segments (left-hand column). We also show (right-hand column) corresponding computed streamlines from numerical work by Higdon [25] in which a high precision, boundary-integral method was used. The physical message of interest is the variation of the trapped eddy pattern within the groove with the ratio $W : h$ of its width (W) to height (h). The three cases shown are (top to bottom) $W : h = 1$, 2, and 3. Considerable agreement between experiment and computation is clearly displayed.

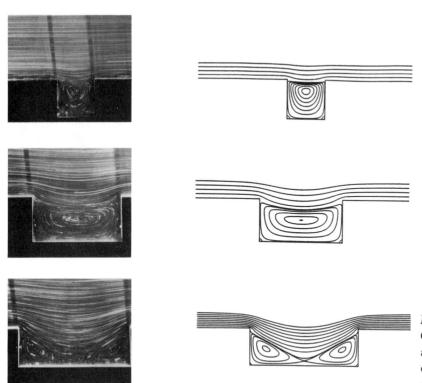

Figure 1.7
Comparisons of experiment (left) and numerical calculations (right) of Stokes flow over rectangular cavities of different aspect ratios.

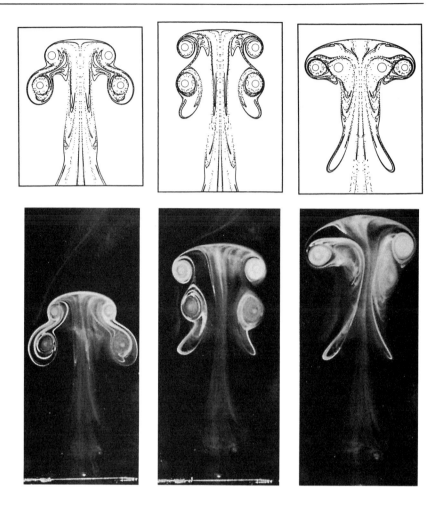

Figure 1.8
Comparison of numerical calculations (top) and experiment (bottom) for the "leapfrogging" of two coaxial vortex rings visualized by a passive tracer.

1.3.2   Leapfrogging Vortex Rings

Two coaxial vortex rings, released one after the other and traveling in the same direction, can engage in an interesting periodic motion where alternately one passes through the other. In the laboratory, three-dimensional instabilities and effects of viscosity end this charming game of "leapfrogging" vortex rings after one or two passes. In a numerical simulation, on the other hand, it can be continued almost indefinitely in the limit of inviscid fluid mechanics. Figure 1.8 displays a comparison at three different times during the evolutionary process of laboratory flow visualizations (bottom) by Yamada [26] and Matsui

[27] with computations (top) by Shariff et al. (1988). While the vorticity distribution in the laboratory visualizations is unknown, the vortices in the computation are tori, small circular cross sections of which are seen in the top three panels within the cloud of marker particles. Considerable similarity in the distribution of the marker is apparent between computation and experiment, strongly suggesting that most of the mushroom shape seen in the experiments is due to dispersion of the smoke and not a breakdown of the vortex rings themselves. This dispersion arises from the fluid motion and not from molecular diffusion, which is, in fact, not included in the computational model. It is believed that chaotic advection is responsible for the general shape of the marker cloud seen in Figure 1.6 (cf. reference [28]).

1.3.3 Colliding Vortex Rings

Constraints of symmetry (and mechanical conservation laws) force the vortices in Figure 1.8 to remain at a finite distance from one another. However, many interactions in vortex flows involve close encounters of vortices with consequent rearrangement of the "topology" of the flow. Fully developed turbulent flows are expected to embody many such events taking place uniformly throughout space (on average) and randomly in time. This view has led to the study of such processes in isolation or in controlled scenarios, a topic sometimes referred to as "synthetic turbulence." A prime example is the collision of vortex rings initially started on intersecting paths. This flow has been studied in several experiments. Those of Oshima and Asaka [29] and Schatzle [30] are particularly noteworthy. In Figure 1.9a-c, we show three 9-panel views from perpendicular directions of the collision and reconnection of two identical vortex rings (one green, one red) and the subsequent emergence of two new rings each made up of half of either original ring. These images are from simulations by I. Zawadzki using (a) a three-dimensional vortex-in-cell code with some 15,000 "vortex particles" making up each ring and (b) an underlying periodic grid with 64^3 nodes. Although we do not show them here, flow visualization images from the laboratory experiments correspond closely to these computational results. The numerical data contain a wealth of information on the trajectories of individual particles within each vortex ring that, hopefully, will lead to an improved understanding of the mechanisms underlying the reconnection process. For further discussion see Aref and Zawadzki [31] and Zawadzki and Aref [32].

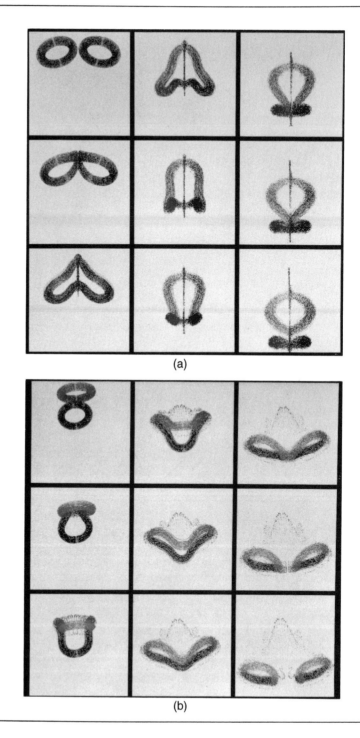

(a)

(b)

Figure 1.9
Numerical simulations of the collision of two vortex rings viewed from three perpendicular directions. The rings are colored red and green initially. Because of the collision and resulting reconnection, two rings, each half green and half red, are produced. See insert for color representation.

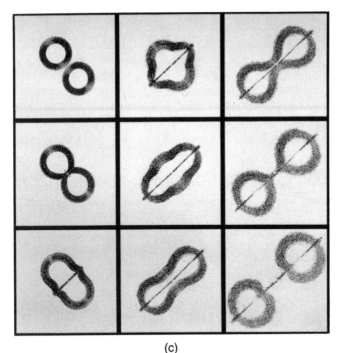

(c)

Figure 1.9
(continued)

1.3.4 Rayleigh-Taylor Instability

In some flows, certain features are immediately singled out for visualization. In flows with sharp interfaces, free surfaces, or shock waves, the shape of that interface, surface, or shock is usually of central importance. While interfaces in two-dimensional flows are just curves and hence easy to plot and visualize, interfaces in fully three dimensional flows are complicated spatial objects that require considerable graphics work to visualize. In Figure 1.10, we show an example from simulations by Tryggvason and Unverdi [33] of Rayleigh-Taylor instability of the interface between two immiscible fluids that results from an initially flat interface when accelerating the fluid of higher density towards the fluid of lower density. Techniques from solids modeling have been used to give the picture perspective and depth, thus facilitating comprehension of the structure. This structure, and its rate of growth, changes with the main control parameters, such as the density difference between the two fluids (relative to the average density).

Figure 1.10
Solid modeling applied to the interface between two immiscible fluids of different density evolving due to Rayleigh-Taylor instability. See insert for color representation.

1.3.5 Free Surface Waves Due to an Oscillating Cylinder

Related to Figure 1.10 in terms of visualization technique, Figure 1.11 shows four frames from a numerical simulation by Chiba and Kuwahara [34] of the wave motion produced at the free surface of a fluid agitated by a cylindrical rod that oscillates back and forth harmonically (and pierces the surface of the fluid). The fluid is initially at rest, and is at rest at infinity. Vortices are shed by the moving cylinder (see panel 1 in Figure 1.11) and are visualized by the free surface waves. The ratio of oscillation amplitude to cylinder diameter is close to 2. The Reynolds number based on oscillation amplitude, frequency, and cylinder diameter is $O(10^4)$. The numerical method is a finite difference method with third-order upwinding and a grid that conforms to the free surface. The computation was carried out on a Hitachi S-820/80 supercomputer using 40 hours of CPU time.

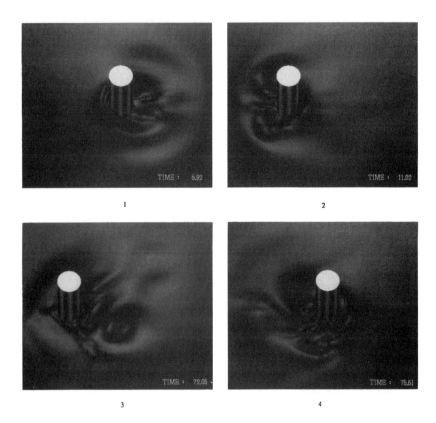

Figure 1.11
Four frames from a video of a numerical simulation of wave motion produced by a harmonically oscillating rod.

Full rendering of the surface and the cylinder are performed, resulting in images of considerable realism. The visualization was done using Phong's shading model on a Fujitsu VP-200 supercomputer. The still images in Figure 1.11 provide only a limited sense of the realism experienced while viewing the full video sequence. This kind of visualization forcefully brings home the message that the equations being solved describe the motion of a fluid. The video has considerable "fluidity."

1.3.6 Laser induced fluorescence (LIF)

There are now many interesting examples of the LIF flow visualization technique in the literature, and we include just one here to illustrate the capabilities of this visualization method. The technique, in general terms, is based upon the principle that one of the components of the flow will fluoresce or "glow" when an excitation or "pumping" beam of light is shown through it. The glow, or "signature," of the fluorescing medium occurs at a wavelength different from the pumping light wavelength and thus can be isolated from the pumping light source by optical filters.

The application shown in Figure 1.12, due to Southerland et al. [35], is aimed at visualizing the molecular mixing in a vortex ring. The figure shows a quantitative visualization of the molecular mixing process in an axisymmetric, laminar vortex ring. A 256×256, 8-bit data plane from a high-resolution, planar, digital imaging measurement of LIF is shown. The concentration field $\xi(\mathbf{x}, t)$ of the ring fluid, denoted by increasing color values from blue through red in the right half of the diametral plane of the ring, is shown in the left panel. In the right panel, the resulting molecular mixing rate field $\nabla \xi \cdot \nabla \xi$ has

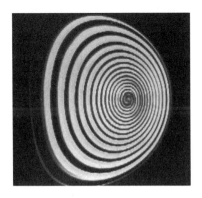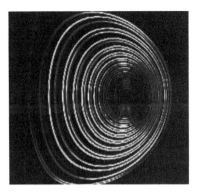

Figure 1.12
LIF visualization of mixing in a vortex ring. See insert for color representation.

been obtained by direct differentiation of the original concentration field data plane, with increasing colors now denoting the local instantaneous value of the mixing rate in the flow. Note that the structure of the mixing process can be followed right to the central core of the ring.

The magnitude of the experiment and visualization effort is typical for modern, computer-augmented flow visualization studies in the laboratory. It is interesting that many of the steps have clear counterparts in modern CFD and its attendant visualization.

1.3.7 Coherent Structures

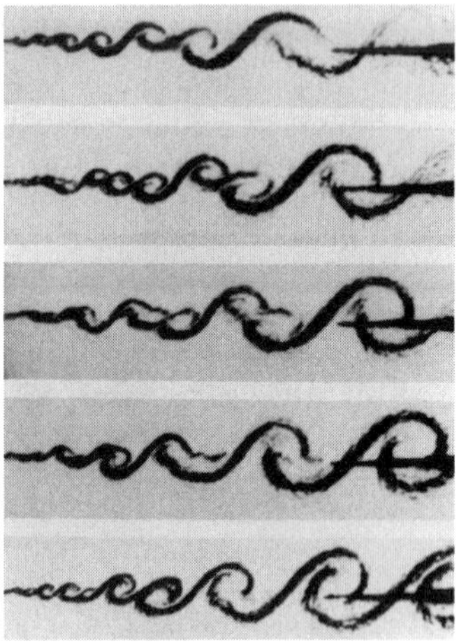

It is imperative to include in our little "gallery" some of the most famous flow visualization pictures, which can justly be said to have altered the course of turbulence research since their appearance in the early 1970s. We are referring, of course, to the discovery of coherent structures in turbulent shear flows, most clearly seen in the mixing layer between parallel streams moving at different speeds. Figure 1.13 shows a sequence of frames from a motion picture by L. Bernal, G. L. Brown, and A. Roshko of the prominent vortices that form in the mixing layer downstream of a splitter plate (off to the left in Figure 1.13 – the protrusion at right is a probe). In this particular sequence, the visualization is due to density differences between the two streams (nitrogen on top and a helium-argon mixture on the bottom). The time sequence (top to bottom panel) shows the merging of two vortices into one (which is at the tip of the probe in the lowest panel). This "vortex pairing" mechanism is another elemental mechanism of great importance in the evolution of turbulent shear flows.

Figure 1.13
Visualization of coherent structures in a mixing layer.

1.3.8 Cosmic Jets

In some subjects where fluid mechanics plays a significant role, such as oceanography, meteorology, and astrophysics, the possibilities for laboratory experimentation are extremely limited. Here computer simulations have proven an invaluable tool for comparing theoretical scenarios with observational data. In planetary atmosphere dynamics, for example, various theories of the formation of large-scale structure and events, such as Jupiter's Great Red Spot, blocking events in the Earth's atmosphere, and El Niño, have been subjected to computer simulation. In Figure 1.14 we show an example of this kind.

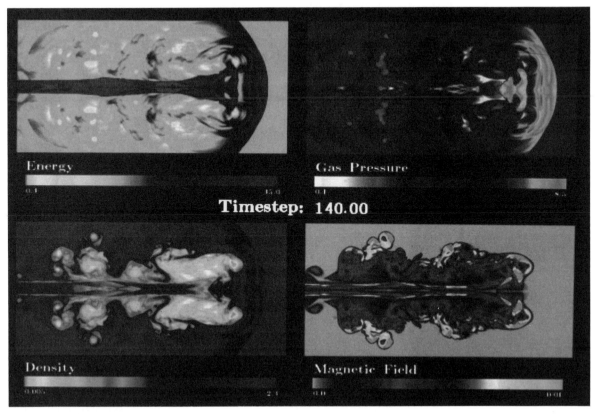

Figure 1.14

False color visualization of fields of energy, pressure, density, and magnetic field in a numerical simulation of a "cosmic jet." See insert for color representation.

Cosmic jets, first observed in the 1950s, appear to be long streamers of hot gases. The gases are expelled from the active centers of galaxies and can extend out as far as hundreds of thousands of light years. Astrophysicists believe that a jet may travel at nearly the speed of light and that its energy may come from a black hole. Some 200 cosmic jets have been identified to date.

The images in Figure 1.14 are from a computer model of a cosmic jet, attempting to reconstruct the jet evolution using the equations of fluid dynamics. The jet is a thin streamer in the middle of the picture. All around it is the backflow, or "cocoon," from the shock wave that forms at the jet's leading edge. The four panels in Figure 1.14 show color coding of the energy, density, gas pressure, and magnetic field in the jet as presented by the computer simulation.

Two outstanding questions are: Why do the jets tend to remain together, collimated, over enormous distances? Why doesn't the backflow just dissipate into the surrounding intergalactic medium? It is hypothesized that magnetic fields generated by internal currents may help to confine the jets. In this model, the gas pressure is stronger than the magnetic field and yet the jet remains relatively well confined. Computer experiments, such as the one shown here, are helping to achieve a better understanding of these mysterious objects.

1.3.9 Flow in a Soap Film

In most cases, use of a continuous color mapping of an experimentally accessible flow field implies the intervention of computer graphics. In Figure 1.15, we show an example where this is emphatically not the case. The jet flow shown here takes place in a thin sheet of fluid trapped within a soap film [36] and the colors are due to interference phenomena of light in the film. Similar visualizations have been used by Couder and collaborators [37,38] and, many years before, by Dewar [39]. An interesting point about these visualizations is that, in all likelihood, they represent the closest approximation to two-dimensional flow possible in a laboratory experiment (at least within a medium that consists of an ordinary fluid). On the other hand, two-dimensional flow simulations are routine in CFD, where the reduction

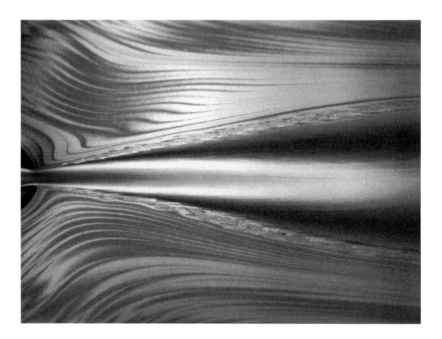

Figure 1.15
Jet flow in a soap film. See insert for color representation.

in dimensionality has for a long time been almost a necessity, but also in many flows a very severe limitation.

1.3.10 Evolution of a Foam

An amusing example of rather peculiar data and flow visualization techniques arises in the evolution of foam. We mention the pioneering studies of three-dimensional foam by Matzke [40,41] and Matzke and Nestler [42], where statistics on thousands of bubbles were collected after photographing each individually and classifying it topologically. In two dimensions, the problem is considerably simpler. In the experiment of Glazier et al. [43] the foam was contained in a flat dish with large, parallel viewing faces. A bit of ink was added to the fluid in the soap films, and data were collected by placing the apparatus on a photocopying machine at regular intervals. Figure 1.16 shows a recent numerical simulation of two-dimensional foam evolution. Aspects of interest are the topological (and, for the purposes of computing stresses, metric) properties of the network of bubbles. In an illustration such as Figure 1.16, due to numerical simulations by T. Herdtle, it is difficult to see where the five-sided bubbles are. On the other hand, if the polygons are color-coded according to the number of sides, this information is readily visible (as in the calculation in Figure 1.17), and the viewer is immediately challenged to ask probing questions concerning geometrical patterns, contiguity of polygons with the same number of sides, and so on. Somewhat analogous uses of color in a vector plot of turbulence data have been promoted by Hunt et al. [44]. For additional information on two-dimensional foam simulations see Herdtle and Aref [45].

1.3.11 Temperature-Sensitive Particles

As our final gallery entry we show in Figure 1.18 an intriguing visualization, due to Dabiri and Gharib [46], in which the particles tracing out the flow are also temperature sensitive and give off light of different colors. (In Figure 1.18 the reddish colors are colder fluid and the the blue-greens warmer fluid!) In this way, the visualization accomplishes two goals at a glance: giving the velocity and the temperature field. Thus, velocity-temperature correlations can be read off at once. The picture also gives a clear indication of the relative extent of the velocity and temperature signals at any time. Implementation of similar visualization in CFD is immediate.

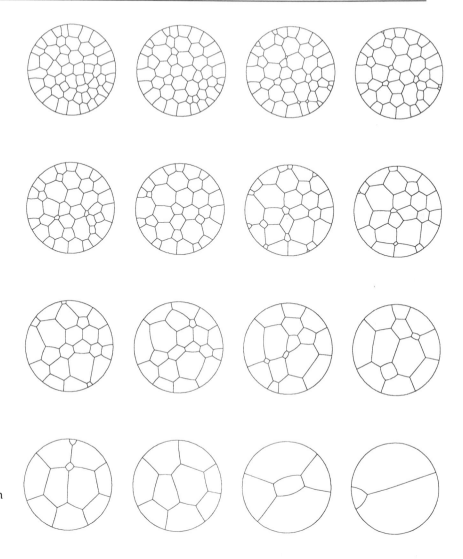

Figure 1.16
Numerical calculation of foam
evolution in a shallow,
cylindrical container.

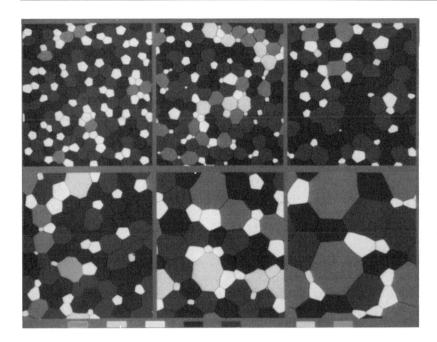

Figure 1.17
Color-coded counterpart to
Figure 1.16. The color coding
immediately reveals the
distribution of bubbles with
different numbers of sides. See
insert for color representation.

Figure 1.18
Simultaneous visualization of
flow velocity and temperature
using liquid crystal particles. See
insert for color representation.

1.4   DISCUSSION

In this section we want to examine some of the apparent issues in flow visualization and identify problems and directions that seem particularly interesting for future research.

The first point to make is that flow visualizations tend to deal with two types of data: *field data* and *particle data*. The fields of interest, in turn, have different tensorial ranks. The most common fields to be visualized in a fluid flow are scalar and vector fields, although a few second-rank tensors are also frequently needed. Particles to be visualized are usually simple, mostly point particles, although particles with an axis[3] occasionally appear (the "vortex elements" in Figure 1.9 are an example).

1.4.1   Displaying Field Data

Several techniques for displaying fields are well established. A scalar field in two dimensions may be visualized using a conventional contour plot. This idea can be generalized to three dimensions by plotting iso-surfaces of the scalar, although one has to face the problem of "looking through" outer surfaces to see inner surfaces. Often iso-surfaces corresponding to just one value of the scalar are shown, with other illustrations reserved for other values. Transparency techniques can be used but result in rather complex illustrations. Another technique is, of course, to use color by associating a range of scalar values with a color bar. This is effective in two dimensions, and sometimes in three, but in cases of overlap of color-coded points such a display becomes confusing. It is a rather viable hypothesis that iso-surfaces in a turbulent flow are, in fact, fractal, so the complexity of the surfaces to be visualized is expected to increase with the value of Reynolds number [47,48].

Vector fields present additional problems because each point in space (or on a computational grid) carries three vector components. In two dimensions, arrow plots can be reasonably effective. In three dimensions, however, perspective is a problem in visualizing arrows: A short arrow seen at right angles may appear longer than a long arrow seen almost head-on, leading to an incorrect visual impression of key qualities of the data.

Vector data in two dimensions can be effectively displayed also by using color coding. A recently developed technique by Charles [49,50] represents the magnitude and direction of a velocity vector by

---

[3] We may call them vector particles.

the intensity and hue, respectively, of a pixel on the color graphics display. The ability to use just one pixel per data point allows a much finer resolution and enhanced visualization of the field.

In the HSV color definition, the letters stand for *hue* (location on a color wheel), *saturation* (how pure the color is), and *value* (how bright a color is). Pure colors are those colors that contain no white, such as a pure red compared to pink, and they lie on the outside surface of the color cone. There is a direct correspondence between the HSV cone and the red, green, and blue signals that color a computer screen display. If you tilt the hollow cone of pure colors so that you are looking down the axis to the dark point, the image appears to be a disk made up of concentric rings of color varying from a black center to a bright outer rim. If all the velocities in a two-dimensional flow field are normalized to the maximum field value, one can map any velocity field to the disk, where orientation of the velocity is denoted by its location on the color wheel, and the magnitude of the velocity determines the brightness (or radius on the disk) of the color. Figure 1.3 provides an example for the case of the vortex street. The regions of rotating flow in the vicinity of the vortices are clearly visible against the background of the main left-to-right flow. In a plot of velocity, in particular, the regions of turning flow stand out as rainbow arcs of color and the stagnation points (i.e., points of vanishing velocity) appear as "black holes" in the overall background color.

1.4.2 Displaying Particle Data

In some computational methods, and in many laboratory visualization techniques, the essential data to be displayed are associated with a system of advected particles. If we consider the paradigm of a photographic recording of moving particles, we arrive at *pathlines* (or *streaklines*) if the aperture is left open for an extended interval of time, but *timelines* if an instantaneous view is realized. CFD visualizations in essence follow the same paradigm, although we now have greater flexibility in color coding of particles according to where they originate, or other telling features.

Particle data are typically not arranged on a grid in space and require techniques from the field of volume visualization for proper display. Because the number of particles is often large, simply plotting a point for each particle may not yield satisfactory results. Sometimes, geometrical structure that is known on physical grounds (e.g., the particles are arranged on curves or surfaces in space) may be useful as an aid to visualization. Vortex lines, or vortex sheets and surfaces across which physical variables such as density, temperature, or viscosity have large jumps, are often helpful in this regard, and there is considerable interest regarding the configuration of such structures.

When displaying particle data, "the relative orderliness of Eulerian representation over Lagrangian," as Amsden and Harlow [51] so aptly put it, must constantly be kept in mind. This phenomenon arises because of chaotic advection of particles.

1.4.3 The Issue of Resolution

There is a physically interesting, and in the long run important, difference between the particles that can be traced in a CFD visualization and the dye or smoke that is routinely traced in laboratory experiments. Dye and smoke appear almost infinitely extensible because of the ratio between macroscopic and molecular scales. The available resolution of smoke or dye thus quickly exhausts the range of any CFD visualization: The number of particles that one would need to follow becomes prohibitively large.

This seems to be a problem area that has not been particularly well addressed and in which interesting algorithms might be found. Generally stated, the problem is a "subgrid scale" modeling problem of how to simulate at lower resolution a phenomenon of higher resolution such that a faithful representation is achieved on the common range of resolved scales. In the dynamical problem of turbulent flow, this kind of problem has led to a considerable literature, which is, in a sense, at the heart of the subject.

The process of rendering appears to compensate for the lack of resolution of the computer-generated image in fluid flow applications as well as in many others. The rendered images of the Rayleigh-Taylor unstable surface (Figure 1.10) or of the free surface supporting a system of waves (Figure 1.11) are not known dynamically at greater resolution than most CFD results. Nevertheless, after rendering the surface appears continuous and to be known to arbitrarily high resolution.

This observation suggests that rendering of dye patches or surfaces can make the viewer think that considerably greater resolution is available than is actually the case. Of course, there is the danger that such an extrapolation can lead to physically unfounded conclusions, but this is not really an issue because one has not forfeited any information by improving the visual sensation of the computer-generated image through rendering and because one can always return to the conventional "low-resolution" image and extract whatever information is available directly from it. The rendering procedure is thus seen as an additional feature of potential usefulness that supplements current conventional visualization procedures.

1.4.4 The Issue of Structure

Visualizing complex three-dimensional objects has always been difficult and, in a fluid flow, is exacerbated by the fact that the structure is continually evolving and typically increasing in complexity with time. In dealing with three-dimensional visualizations, it seems to be essential to be able to rotate and zoom with ease so that the best view of a given structure can be obtained.

Wire-frame images are a first step in visualizing three-dimensional objects but are quickly becoming unsatisfactory. Tools from solids modeling are now being used to bring out the structure in greater detail. Apart from Figures 1.10 and 1.11, we refer the reader to the recent work of Helman and Hesselink [52]. The merits of these techniques is that they appeal to the processing used by the human eye and brain.

An intriguing option for capturing three-dimensional structure in a single image is the production of a computer-generated hologram. Computer-generated holography (CGH) is the process of computing the pattern needed to holographically represent stored data and then transferring that pattern directly onto a recording material. In CGH there is no need for a physical object or for a precisely aligned coherent optical system to do the recording. The hologram is a completely portable hardcopy containing all three-dimensional views.

Standard two-dimensional representations of three-dimensional data have some well-known shortcomings. Among them, each different view of the data must be recalculated so it is often difficult to preview a series of views to see the relative parallax motion of parts of an image at different depths. Two-dimensional projections also fail to engage most of the three-dimensional vision capabilities of the observer.

Various experimental groups are now using holograms for the representation of data. The recent review by Hesselink [53] includes a hologram. It follows that the corresponding groups doing computer simulation will want to have purely digital tools available to accomplish similar tasks.

Medium-resolution holograms can be recorded using standard film recorders that are already in use for production of high-resolution two-dimensional images. Higher-resolution holograms can be recorded on electron beam lithography machines, systems which are currently used to etch prototype integrated circuits.

1.4.5 Computer Models of Laboratory Visualizing Agents

In recent years, entirely "synthetic" images of considerable realism have been generated of clouds [54], water waves [55,57], and smoke [58]. These images use some kind of fractal construction, either through the use of iterated function systems or through fractional Brownian motion. There is essentially no calculation of fluid dynamical phenomena required to produce such images. However, these developments do suggest that it is possible to write improved algorithms for tracing smoke and/or dye that would result in more realistic images to be compared quite directly to images generated in the laboratory and recorded photographically. Until now, the common ground between an experimental investigation and a CFD calculation has typically been a contour plot or some other "quantitative" comparison of this kind. A rendered computer image that attempts to approach a photograph of a flow visualization from the laboratory may be equally quantitative and may in fact ensure a greater level of global agreement. The numerical modeling of the optics of smoke also introduces fundamental physics issues of considerable interest [56].

1.5 CONCLUSIONS

Representation of data in graphical form is a practice that pervades much of fluid mechanics, both experimental and computational. It is natural to reflect on why this practice demands so much interest and whether the effort spent on visualization is worthwhile from a scientific point of view.

In the laboratory, visualization plays the natural role of recording, for analysis and for presentation to others, of the experimental "evidence" of some phenomenon. In CFD the issue of validating the graphics generating code arises just as well as the validation of the flow simulation code. To be acceptable, visualization must be "valid." The visualizer must avoid bias introduced by peculiar uses of color or other effects. Certain standards for what constitutes good visualization need to be established. For example, it has become a convention that in a flow visualization picture the flow is from left to right unless something else is explicitly stated. Contour levels that are represented by colored regions must follow a color bar that follows the optical spectrum. Such conventions and standards will become ever more important as visualization is automated and "push-button" applications become commonplace.

There is a clear division between computer graphics and scientific visualization: the former is a tool of the latter, but a tool that can be used for several other tasks as well. The essential result of performing a scientific visualization is to gain insight into the science being investigated. Visualization initiates or facilitates a "feedback loop" where a scientific calculation through visualization leads to new questions to be asked of the science, which in turn lead to new visualizations, and so forth. Such a "loop" is absent for general graphics, at least as far as scientific insight is concerned. Visualization that simply results in a set of pretty pictures which do not give rise to additional understanding and analysis has in this view failed.

The richness and beauty of scientific visualization images in fluid mechanics is sure to increase, as it has for the past decade. As workstations become faster and more versatile, and scientific users learn more about the available graphics software (and maybe develop some of their own), images are going to increase in detail, information content, and aesthetic appeal.

ACKNOWLEDGMENTS

We thank S. Chiba, W. J. A. Dahm, M. Gharib, T. Herdtle, S. W. Jones, G. Tryggvason, and I. Zawadzki for assistance with the illustrations. H. A. acknowledges research support from NSF/PYI award CTS84-51107 and DARPA/ACMP URI grant N00014-86-K-0758.

REFERENCES

[1] H.E. Edgerton and J.R. Killian, Jr., *Moments of Vision*, The MIT Press, Cambridge, MA (1979).

[2] A.M. Worthington, *A Study of Splashes*, Longmans Green, New York (1908).

[3] H. Reichenbach, Contributions of Ernst Mach to Fluid Mechanics, *Annu. Rev. Fluid Mech.* **16**, 99-137 (1983).

[4] D.J. Tritton, *Physical Fluid Dynamics*, second edition, Clarendon Press, Oxford (1988).

[5] H. Werlé, Hydrodynamic Flow Visualization, *Annu. Rev. Fluid Mech.* **5**, 361-382 (1973).

[6] T. Asanuma (ed.), *Flow Visualization*, Hemisphere; McGraw-Hill, New York (1979).

[7] W. Merzkirch, *Flow Visualization*, second edition, Academic Press, New York (1987).

[8] S.J. Kline, Flow Visualization (Motion Picture), National Committee on Fluid Mechanics Films; Distributed by Encyclopaedia Brittanica Film Center, Chicago (1963).

[9] M. Van Dyke, *An Album of Fluid Motion*, Parabolic Press, Stanford (1982).

[10] H. Reed, Gallery of Fluid Motion, *Phys. Fluids* **28**, 2631-2640 (1985).

[11] H. Reed, Gallery of Fluid Motion, *Phys. Fluids* **29**, 2769-2780 (1986).

[12] H. Reed, Gallery of Fluid Motion, *Phys. Fluids* **30**, 2597-2606 (1987).

[13] H. Reed, Gallery of Fluid Motion, *Phys. Fluids* **31**, 2383-2394 (1988).

[14] H. Reed, Gallery of Fluid Motion, *Phys. Fluids A* **1**, 1439-1450 (1989).

[15] H. Reed, Gallery of Fluid Motion, *Phys. Fluids A* **2**, 1517-1527 (1990).

[16] S. Corrsin and M.J. Karweit, *J. Fluid Mech.*, **39**, 87-96 (1969).

[17] H. Aref, Stirring by Chaotic Advection, *J. Fluid Mech.* **143**, 1-21 (1984).

[18] H. Aref, S.W. Jones, and O.M. Thomas, Computing Particle Motions in Fluid Flows, *Comput. in Physics* **2**(6), 22-27 (1988).

[19] H. Aref, S.W. Jones, S. Mofina, and I. Zawadzki, Vortices, Kinematics and Chaos, *Physica D* **37**, 423-440 (1989).

[20] V.I. Arnol'd, Sur la Topologie des Écoulements Stationnaires des Fluides Parfaits, *C. R. Acad. Sci. Paris A* **261**, 17-20 (1965).

[21] M. Hénon, Sur la Topologie des Lignes Courant dans un Cas Particulier, *C. R. Acad. Sci. Paris A* **262**, 312-314 (1966).

[22] T. Dombre, V. Frisch, J.M. Greene, M. Hénon, A. Mehr, and A.M. Soward, Chaotic Streamlines and Lagrangian Turbulence: The ABC Flows, *J. Fluid Mech.* **167**, 353-391 (1986).

[23] H. Helmholtz, On Integrals of the Hydrodynamic Equations which Express Vortex-Motion, translation by P. G. Tait in *Phil. Mag.* **33**(4), 485-512 (1867).

[24] S. Taneda, Visualization of Separating Stokes Flows, *J. Phys. Soc. Japan* **46**, 1935-1942 (1979).

[25] J.J.L. Higdon, Stokes Flow in Arbitrary Two-Dimensional Domains: Shear Flow Over Ridges and Cavities, *J. Fluid Mech.* **159**, 195-226 (1985).

[26] H. Yamada and T. Matsui, *Phys. Fluids*, **21**, 292-294 (1978).

[27] K. Shariff and A. Leonard, *Ann. Rev. Fluid Mech.*, **24**, 235-279 (1992)

[28] V. Rom-Kedar, A. Leonard, and S. Wiggins, An Analytical Study of Transport, Mixing and Chaos in an Unsteady Vortical Flow, *J. Fluid Mech.* **214**, 347-394 (1990).

[29] Y. Oshima and S. Asaka, Interaction of Two Vortex Rings Moving Side by Side, *Nat. Sci. Rep. Ochanomizu Univ.* **26**, 31-37 (1975).

[30] P. Schatzle, *An Experimental Study of Fusion of Vortex Rings*, Ph.D. Thesis. Graduate Aeronautical Laboratories, California Institute of Technology (1987).

[31] H. Aref and I. Zawadzki, Linking of Vortex Rings, *Nature* **354**, 50-53 (1991).

[32] I. Zawadzki and H. Aref, Mixing During Vortex Ring Collisions, *Phys. Fluids A* **3**, 1405-1410 (1991).

[33] G. Tryggvason and S.O. Unverdi, Computations of Three-Dimensional Rayleigh-Taylor Instability, *Phys. Fluids A* **2**, 656-658 (1990).

[34] S. Chiba and K. Kuwahara, Numerical Analysis of Free Surface Flow Around a Vertical Cylinder, *Bull. Am. Phys. Soc.* **34**, 2259 (1989).

[35] K.B. Southerland, J.R. Porter, W.J.A. Dahm, and K.A. Buch, An Experimental Study of the Molecular Mixing Process in an Axisymmetric, Laminar Vortex Ring, *Phys. Fluids A* **3**, 1385-1392 (1991).

[36] M. Gharib and P. Derango, A Liquid Film (Soap Film) Tunnel to Study Two-Dimensional Laminar and Turbulent Shear Flows, *Physica D* **37**, 406-416 (1989).

[37] Y. Couder and C. Basdevant, Experimental and Numerical Study of Vortex Couples in Two-Dimensional Flows, *J. Fluid Mech.* **173**, 225-251 (1986).

[38] Y. Couder, J.M. Chomaz, and M. Rabaud, On the Hydrodynamics of Soap Films, *Physica D* **37**, 384-405 (1989).

[39] J. Dewar, Soap Films as Detectors: Stream Lines and Sound, *Proc. Royal Inst.* **24**, 197-259 (1923). (Also in Lady Dewar (ed.), *Collected Papers of Sir James Dewar*, Cambridge University Press, Cambridge, England, pp. 1334-1379 (1927).

[40] E.B. Matzke, The Three-Dimensional Shapes of Bubbles in Foams, *Proc. Natl. Acad. Sci. USA* **31**, 281-289 (1945).

[41] E.B. Matzke, The Three-Dimensional Shape of Bubbles in Foam – An Analysis of the Role of Surface Forces in Three-Dimensional Cell Shape Determination, *Am. J. Bot.* **33**, 58-80 (1946).

[42] E.B. Matzke and J. Nestler, Volume-Shape Relationships in Variant Foams. A Further Study of the Role of Surface Forces in Three-Dimensional Cell Shape Determination, *Am. J. Bot.* **33**, 130-144 (1946).

[43] J.A. Glazier, S.P. Gross, and J. Stavans, Dynamics of Two-Dimensional Soap Froths, *Phys. Rev. A* **36**, 306-312 (1987).

[44] J.C.R. Hunt, A.A. Wray, and P. Moin, Eddies, Streams, and Convergence Zones in Turbulent Flows, in *Center for Turbulence Research Rep. CTR-S88*, pp. 193-208 (1988).

[45] H. Aref and T. Herdtle, *J. Fluid Mech.*, **241**, 233-260 (1993).

[46] D. Dabiri and M. Gharib, Digital Particle Image Thermometry – The Method and the Implementation, *Exp. in Fluids* **11**, 77-96 (1991).

[47] B.B. Mandelbrot, On the Geometry of Homogeneous Turbulence, with Stress on the Fractal Dimension of the Iso-Surfaces of Scalars, *J. Fluid Mech.* **72**, 410-416 (1975).

[48] K.R. Sreenivasan and C. Meneveau, The Fractal Facets of Turbulence, *J. Fluid Mech.* **173**, 357-386 (1986).

[49] R.D. Charles, A New Color Graphics Display Technique for Flow Velocity Information, *Gather/Scatter* **5**, 10-11 (1989).

[50] R.D. Charles, Viewing Velocity in Flow Fields, *Mech. Eng.* **111**(8), 64 (1989).

[51] A.A. Amsden and F.H. Harlow, Slip Instability, *Phys. Fluids* **7**, 327-334 (1964).

[52] J. Helman and L. Hesselink, Analysis and Visualization of Flow Topology in Numerical Data Sets, in *Topological Fluid Mechanics*, H. K. Moffatt and A. Tsinober (eds.), Cambridge University Press, New York, pp. 361-371 and color plates (1990).

[53] L. Hesselink, Digital Image Processing in Flow Visualization, *Annu. Rev. Fluid Mech.* **20**, 421-485 (1988).

[54] G.V. Gardner, Visual Simulation of Clouds, Computer Graphics **19**(3), 297-303 (1985).

[55] D.R. Peachey, Modeling Waves and Surf, *Comput. Graphics* **20**(4), 65-74 (1986).

[56] M.V. Berry, and I.C. Percival, Optics of Fractal Clusters Such as Smoke, *Opt. Acta* **33**, 577-591 (1986).

[57] A. Fournier and W.T. Reeves, A Simple Model of Ocean Waves, *Comput. Graphics* **20**(4), 75-84 (1986).

[58] M.F. Barnsley, Fractal Modelling of Real World Images, in *The Science of Fractal Images*, H-O. Peitgen and D. Saupe (eds.), Springer-Verlag, New York, pp. 219-242 (1988).

[59] L. Prandtl and O.G. Tietjens, *Applied Hydro- and Aeromechanics*, McGraw-Hill, New York (1934). [Transl. from the German edition, Springer, Berlin (1931)].]

2

Visualization of Scroll Waves

Mario Markus and Manfred Krafczyk

Max-Planck-Institut für Ernährungsphysiologie
Dortmund, Germany

2.1 INTRODUCTION

In this chapter, we report on simulations and graphical displays of beautiful and intricate three-dimensional waves called "scroll waves" in special kinds of media. Let's summarize the main features of these waves and media before proceeding further. In particular, we are fascinated by "excitable media." In each point of space, an excitable medium can undergo a fast transition into an "excited" state if triggered by a threshold-transgressing perturbation coming from the neighborhood of the point. For details on this type of media, the reader may consult references 1–3. After this excitation, the medium becomes "refractory," slowly recovering its excitability until it becomes "receptive" and can be excited again. In this way, waves may be formed which have the following properties: (i) They suffer no attenuation, (ii) high-frequency waves annihilate low-frequency waves, (iii) waves with equal frequencies annihilate each other, and (iv) waves are not reflected at the boundaries of the medium. This type of waves appears in the entire spectrum of macroscopic science. Examples are found in nerve axons [4], the retina [5], fertilized eggs [6], the cerebral cortex around electrically stimulated points and epileptic foci [7], heart tissue [8–12], aggregating populations of the so-called "social amoeba" (slime mold *D. discoideum*) [13–16], the spreading of infectious diseases [17–19], oxidation processes on metal surfaces [20], the Belousov-Zhabotinskii (BZ) reaction [21–30] and spiral galaxies [24,31,32].

The simulations presented here are performed using a cellular automaton. Cellular automata are a class of simple mathematical systems which are becoming important as models for a variety of physical processes. They are mathematical idealizations of physical systems in which space and time are discrete. They can exhibit random-looking behavior sometimes and highly ordered behavior at others – depending on the rules of the game. Interesting previous applications include the modeling of the spread of plant species, the propagation of animals such as barnacles, and the spread of forest fires. Usually cellular automata consist of a grid of cells which can exist in two states, occupied or unoccupied. The occupancy of one cell is determined from a simple mathematical analysis of the occupancy of neighbor cells. Mathematicians define the rules, set up the game board, and let the game play itself out. One popular set of rules is set forth in what has become known as the game of LIFE. Though the rules governing the creation of cellular automata are simple, the patterns they produce are very complicated and sometimes seem almost random, like a turbulent fluid flow or the output of a cryptographic system.

Cellular automata have proven to be efficient alternatives to the integration of partial differential equations in different areas of science (see, e.g., references 33–37). However, automata simulating excitable media with a periodic lattice suffer from the drawback that the shape of the cells (squares [36,38], hexagons [24], or cubes [39]) propagates into macroscopic scales, leading to waves of the same (or related) shape. Thus, the wave propagation is anisotropic. In previous publications [40–44], we have developed an isotropic automaton by assuming a semi-random distribution of excitable elements. In these works, the automaton was used mainly to simulate two-dimensional waves, yielding quantitative agreement with experiments in the BZ reaction given in reference 30.

In the present work, we deal with a three-dimensional generalization of our isotropic semi-random automaton and concentrate on techniques for the visualization of different types of the resulting scroll waves. A number of theoretical works on scroll waves using alternative methods have been performed [45–51]. Unfortunately, only very few experimental observations (in thick heart muscle [12], as well as in the BZ reaction in solution [21,23] and in gels [29,52,53]) have been reported.

One of the intentions of this chapter is to propose visualization techniques to the theoreticians. Furthermore, we intend to facilitate the interpretation of experimental results by pointing to different types of wave shapes that may develop and to the ways they may appear to the observer, depending on the viewing technique or the viewpoint.

2.2 THE MODEL

Excitable media can be characterized by equations of the type

$$\frac{\partial A}{\partial t} = f(A, B) + D_A \Delta A \tag{2.1}$$

$$\frac{\partial B}{\partial t} = g(A, B) + D_B \Delta B \tag{2.2}$$

(see, for example, references 2, 26, 46, and 50). Let's explain these equations. Sometimes it helps to first consider their general significance in the field of chemistry where A and B denote the concentrations of reactants A and B. The left side of the equations represents the change in concentrations through time. D represents a diffusion coefficient. The ΔA and ΔB terms relate to the diffusion or dissipation of the chemical species. The f and g terms are reactant terms which are functions of the concentration of A and B. Typical "nullclines," where $f(A, B) = 0$ and $g(A, B) = 0$, as well as trajectories on the A-B plane are shown in Figure 2.1.

We discretize time as follows: The transition from the receptive state (α) to the excited state (β) requires one iteration. The reverse

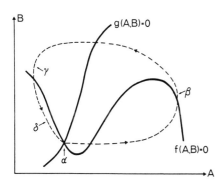

Figure 2.1
Phase-plane diagram for an excitable medium, showing the local dynamics of an excitation variable A and a recovery variable B (see reference 2). α: receptive state. β: excited state. γ, δ: refractory states. Solid lines: nullclines. Dashed curve: trajectory obtained by excitation of the receptive state.

Figure 2.2
Geometry of our
three-dimensional cellular
automaton. The medium is
divided into cubic cells of side
length d. One excitable point is
placed randomly in each cell, as
illustrated at the left. A typical
distribution of points in a grid
with 6 × 6 × 5 cells is shown at
the right (the cells are not
shown). The neighborhood of
each point is defined by a
sphere with radius R. We
define $r = R/d$.

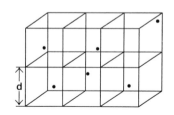

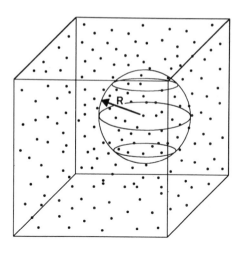

transition usually takes much longer (see, for example, references 54
and 55) and is given in our model by $n + 1$ iterations, where n (the
number of refractory states) is a model parameter. For the sake of
simplicity, we do not take into account in the present chapter the
possibility that refractory states close to α (such as δ) may also become
excited (see reference 43).

Figure 2.2 shows the geometry of our three-dimensional cellular
automaton. The medium under consideration is divided into cubes
of side length d. One excitable point is placed randomly in each cell
at $t = 0$, as exemplified for six cells at the left of Figure 2.2, and this
distribution remains the same throughout the simulation. The right
part of Figure 2.2 shows a distribution of points for a medium divided
into 6 × 6 × 5 cells. For better visualization, the cubes are not shown at
the right part of Figure 2.2. This semi-random distribution of one point
placed randomly in each cell, instead of a fully random distribution of
each point within the whole medium, prevents large inhomogeneities
while keeping the number of points low. A state S is assigned to
each point: $S = 0$ (receptive state), $S = n + 1$ (excited state), and
$S = n, n - 1, \ldots, 1$ (refractory states). At each point, $S(> 0)$ decreases
by one at each iteration step. The excitation of a point is triggered
by the points in its neighborhood, which is defined here by a sphere
with radius R, as illustrated in Figure 2.2. A point becomes excited if
the number ν of excited points inside the spherical neighborhood is
larger than a given threshold m.

Before determining a state $S(t + \tau)$ from a state $S(t)$ (τ is the time
step per iteration), we determine an intermediate state $\sigma(t)$ which

thereafter yields $S(t + \tau)$ by averaging over the neighborhood. The intermediate state $\sigma(\tau)$ is determined by the following rules:

1. *Excitation process:* If $S(t) = 0$ and $\nu \geq m$, then $\sigma(t) = n + 1$.

2. *Stationarity of the receptive states:* If $S(t) = 0$ and $\nu < m$, then $\sigma(t) = 0$.

3. *Loss of excitation and refractoriness in time:* $\sigma(t) = S(t) - 1$ if $S(t)$ is not equal zero.

The new state $S(t + \tau)$ is determined as the next integer to the average of $\sigma(t)$ over all points inside the spherical neighborhood.

This process of "local averaging" takes into account transport processes as in reference 56. We did not perform this averaging for $\sigma = 0, 1, n + 1$; otherwise we obtained numerical instabilities in wavefronts and wavebacks. Altogether we have three control parameters: $r = R/d$, n, and m. In the examples shown in the present work, we set $r = 3$, $n = 3$, and $m = 1$. In all calculations, we use a medium consisting $200 \times 200 \times 200$ cells. Visualization techniques, however, are only applied to subsets of the medium where interesting phenomena take place.

2.3 VISUALIZATION TECHNIQUES

We introduce three visualization techniques: cylindrical slicing, planar slicing, and planar sectioning. For using these techniques, we define three directions as given in Figure 2.3: **a** ("from above"), **s** ("from the side"), and **f** ("from the front").

We explain the technique of cylindrical slicing with the help of Figure 2.3. After choosing the direction **a** as cylinder axis, we draw a cylinder C_1 with radius R_1 and a cylinder C_2 with radius R_2 ($R_2 < R_1$). We define $r_1 = R_1/d$ and $r_2 = R_2/d$. In the graphical representation, we show all the points which are contained in the cylindrical ring C between the curved surfaces of C_1 and C_2. For this sake, we cut the ring with a rectangle T, as shown in Figure 2.3. We display a right parallelepiped P which is constructed as follows. The front F_1 of P results from cutting the curved surface of C_1 by T. The back F_2 of P results from cutting the curved surface of C_2 by T and stretching the resulting surface so that F_2 has the same size and shape as F_1. Accordingly, the points in the solid C are rearranged in the solid P. Interpolations are then performed in order to fill the gaps caused by the stretching process in the back part of C, so as to obtain an ap-

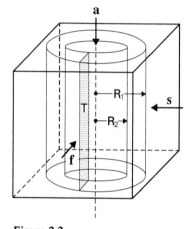

Figure 2.3
Directions defined by vectors **a** ("from above"), **s** ("from the side") and **f** ("from the front"). After choosing a cylinder axis, R_1 and R_2 define a cylindrical ring. This ring is cut by the rectangle T and bent so as to yield a right parallelepiped. This parallelepiped is displayed when using the visualization technique called *cylindrical slicing*. We define $r_1 = R_1/d$ and $r_2 = R_2/d$.

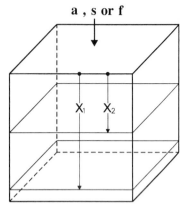

a , s or f

X_1 X_2

Figure 2.4

Explanation of the technique called *planar slicing*. After choosing one of the vectors **a**, **s**, or **f** (see caption of Figure 2.3) as the viewing direction, the distances X_1 and X_2 define a right parallelepiped, which is displayed. We define $x_1 = X_1/d$ and $x_2 = X_2/d$.

proximately constant density of points all over P. In this graphical representation, only the excited points ($S = n + 1$) are displayed. The points at the front surface F_1 are shown in white, and the gray level gets darker as the points lay deeper (i.e., closer to the back surface F_2).

We explain the technique of planar slicing with help of Figure 2.4. After selecting one of the vectors **a**, **s**, or **f** as the viewing direction, the medium is cut by two planes, P_1 resp. P_2, perpendicular to the viewing direction. These planes are defined by the distances X_1 resp. X_2, as indicated in Figure 2.4. We define $x_1 = X_1/d$ and $x_2 = X_2/d$. In the graphical representation, only the excited points ($S = n+1$) lying in the right parallelepiped between P_2 (front) and P_1 (back) are displayed. The points at the front surface P_2 are shown in white, and the gray level gets darker as the points lay deeper (i.e.,closer to the back surface P_1).

The technique of planar sectioning is analogous to that of planar slicing (see Figure 2.4), except that only one layer of cells is displayed (i.e., $X_2 = X_1 - d$). Furthermore, not only the excited points, but also all points in the layer, are shown. The excited points ($S = n + 1$) are shown in white, and the gray level gets darker on decreasing S. Receptive points are shown in black.

2.4 INITIAL CONDITIONS

The initial conditions in the present work are based on the construction of the following ribbons: an untwisted, unknotted ribbon R_{UU}^1 (Figure 2.5a), a ribbon R_T^1 twisted by 360° (Figure 2.5b), and a knotted ribbon R_K^1 (Figure 2.5c). In addition, we consider a ribbon R_{2T}^1 consisting of two halves, one twisted clockwise and the other counterclockwise by 360° (Figure 2.5d).

In a strict sense, we use the word "ribbon" for a two-dimensional manifold embedded in three-dimensional space which, in the cases considered here, has an inner and an outer surface. In order to establish the initial conditions for the computations, we define "adjacent ribbons" R_j^i, where $j = UU, T, K$, or $2T$, $i = 2, \ldots, n + 2$. The R_j^i have the same area and shape as R_j^1, but are displayed perpendicularly a distance $i \cdot R$ in the direction from the inner to the outer ribbon surface. The surface pairs (R_j^{i+1}, R_j^i), $i = 1, 2, \cdots, n + 1$, in addition to the perpendicular surfaces containing the edges of R_j^{i+1} and R_j^i, define solids S_j^i. In Figure 2.6, we show S_j^{n+1}, $j = T, K, 2T$, using the

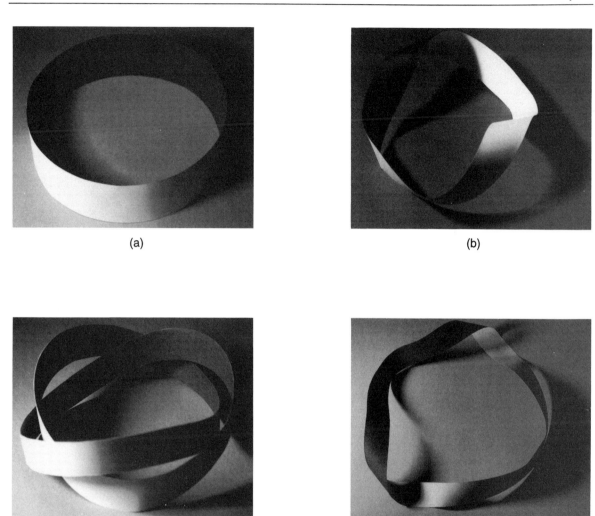

(a)

(b)

(c)

(d)

Figure 2.5
Topology of ribbons for the construction of the initial conditions. (a) Untwisted, unknotted ribbon; (b) ribbon twisted by 360°; (c) knotted ribbon; (d)
ribbon with two halves, one twisted clockwise and the other counterclockwise
by 360°.

Figure 2.6
Initial conditions used in the present work, displayed by the technique of cylindrical slicing. $r_1 = 99$ and $r_2 = 30$. Only the excited state is shown. The conditions correspond to a ribbon twisted by 360° (a), a knotted ribbon (b), and a ribbon consisting of two halves, one twisted clockwise and the other counterclockwise by 360° (c).

(a)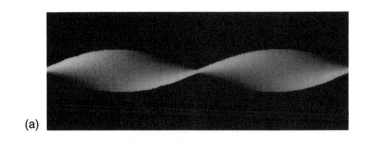

(b)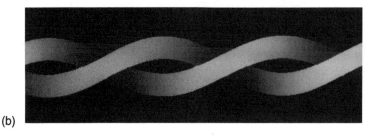

(c)

Figure 2.7
Development of one edge of an untwisted, unknotted scroll wave after 16 iterations.

graphical technique of cylindrical slicing. The initial condition I_j for the calculations in this work are set as follows: All automaton points contained in S_j^i ($i = 1, 2, \ldots n+1$) are in the state $S = i$; All other points are in the receptive state $S = 0$.

2.5 RESULTS

We call $W_j(k)$ the scroll wave obtained from the initial condition I_j ($j = UU, T, K, 2T$) after k iterations – that is, after the time $k\tau$. Figure 2.7 shows $W_{UU}(16)$. This scroll wave is obtained from the simplest initial configuration considered here, namely that corresponding to an untwisted, unknotted ribbon such as that in Figure 2.5a.

Figure 2.8 shows different cylindrical slicings of $W_T(24)$. The ini-

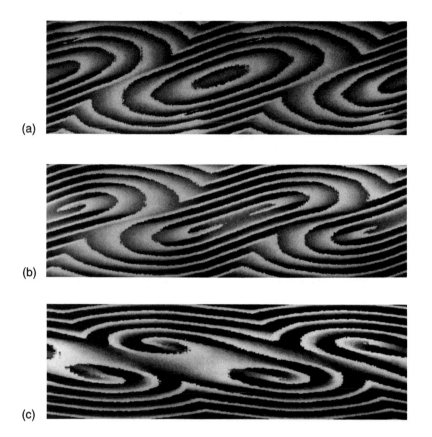

(a)

(b)

(c)

Figure 2.8
Scroll wave resulting from a ribbon twisted by 360° after 24 iterations. The initial conditions correspond to Figure 2.6a. Visualization technique: cylindrical slicing. (a) $r_1 = 99$, $r_2 = 59$; (b) $r_1 = 80$, $r_2 = 40$; (c) $r_1 = 31$, $r_2 = 11$.

tial conditions correspond to a ribbon twisted by 360° similar to that in Figure 2.5b. A cylindrical slicing of the initial conditions is given in Figure 2.6a. Figures 2.9 through 2.11 show $W_T(24)$ using the techniques of (a) planar slicing and (b) planar sectioning for the viewing directions **f** (Figure 2.9), **s** (Figure2.10), and **a** (Figure 2.11).

Figure 2.12 shows different cylindrical slicings of $W_K(19)$. The initial conditions correspond to a knotted ribbon, similar to that in Figure 2.5c. A cylindrical slicing of the initial conditions is given in Figure 2.6b. Figures 2.13 through 2.15 show $W_K(19)$ using the techniques of (a) of planar slicing and (b) planar sectioning for the viewing directions **f** (Figure 2.13), **s** (Figure 2.14) and **a** (Figure 2.15).

Figure 2.16 shows different cylindrical slicings of $W_{2T}(13)$. The initial conditions correspond to a ribbon consisting of two halves, one twisted clockwise and the other counterclockwise by 360° (similar to

Figure 2.9
Same as Figure 2.8, but (a) using the visualization technique of planar slicing with $x_1 = 100$ and $x_2 = 200$ and (b) using the technique of planar sectioning with $x_1 = 100$. Viewing direction: **f**.

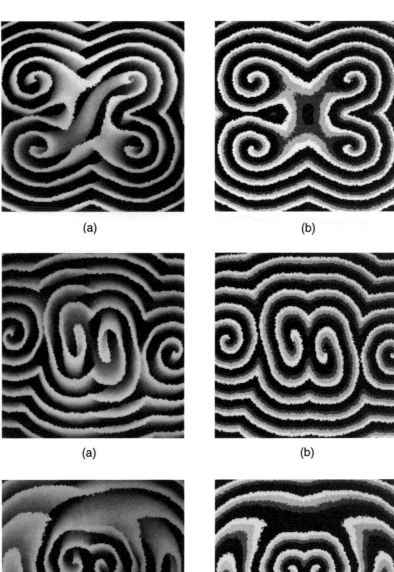

(a)

(b)

Figure 2.10
Same as Figure 2.9, but viewing direction **s**.

(a)

(b)

Figure 2.11
Same as Figure 2.9, but viewing direction **a**.

(a)

(b)

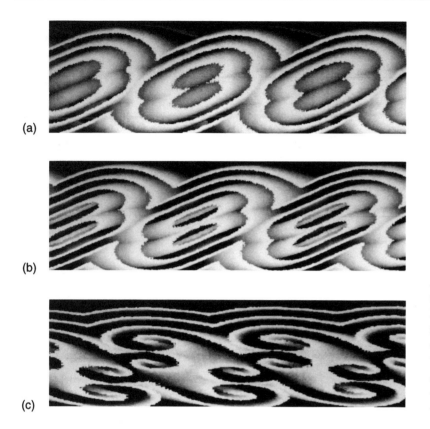

(a)

(b)

(c)

Figure 2.12
Scroll wave resulting from a knotted ribbon after 19 iterations. The initial conditions correspond to Figure 2.6. Visualization technique: cylindrical slicing. (a) $r_1 = 99$, $r_2 = 60$; (b) $r_1 = 60$, $r_2 = 40$; (c) $r_1 = 31$, $r_2 = 10$.

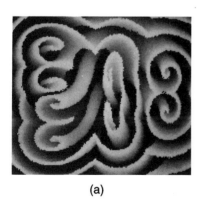

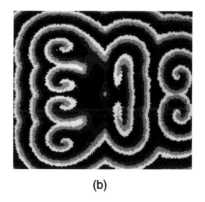

(a)   (b)

Figure 2.13
Same as Figure 2.12, but (a) using the visualization technique of planar slicing with $x_1 = 100$ and $x_2 = 200$ and (b) using the technique of planar sectioning with $x_1 = 100$. Viewing direction: **f**.

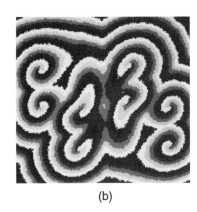

(a)   (b)

Figure 2.14
Same as Figure 2.13, but viewing direction **s**.

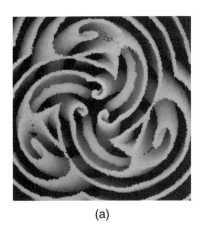

(a)   (b)

Figure 2.15
Same as Figure 2.13, but viewing direction **a**.

that in Figure 2.5d). A cylindrical slicing of the initial conditions is given in Figure 2.6c. Figure 2.17 through 2.19 show $W_{2T}(13)$ using the techniques of (a) planar slicing and (b) planar sectioning for the viewing directions **f** (Figure 2.17), **s** (Figure 2.18) and **a** (Figure 2.19). For comparison with the plane sectioning in Figure 2.17b, which does not reveal the asymmetry between the upper and the lower halves of the wave, an additional planar sectioning for the viewing direction **f** is shown in Figure 2.17c.

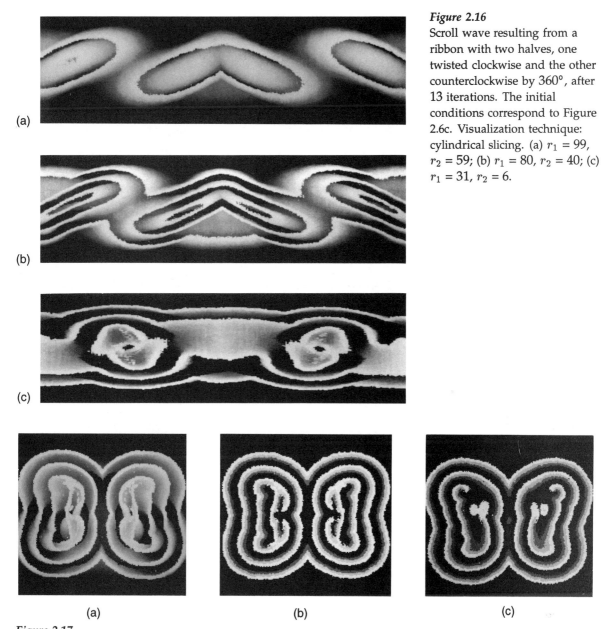

Figure 2.16
Scroll wave resulting from a ribbon with two halves, one twisted clockwise and the other counterclockwise by 360°, after 13 iterations. The initial conditions correspond to Figure 2.6c. Visualization technique: cylindrical slicing. (a) $r_1 = 99$, $r_2 = 59$; (b) $r_1 = 80$, $r_2 = 40$; (c) $r_1 = 31$, $r_2 = 6$.

(a)

(b)

(c)

(a) (b) (c)

Figure 2.17
Same as Figure 2.16 but (a) using the visualization technique of planar slicing with $x_1 = 100$ and $x_2 = 200$ and (b,c) using the technique of planar sectioning with $x_1 = 100$ (b) and $x_1 = 107$ (c). Viewing direction: **f**

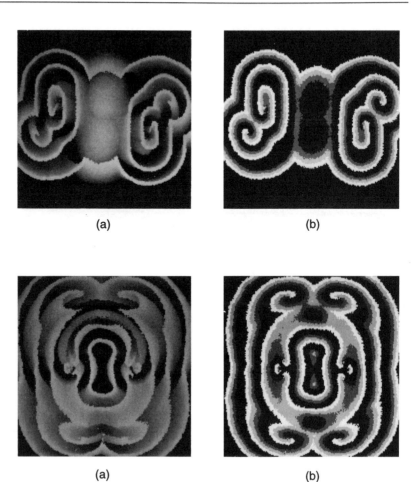

Figure 2.18
Same as Figure 2.17 but using planar slicing with $x_1 = 103$, $x_2 = 200$ (a) and using planar sectioning with $x_1 = 103$ (b). Viewing direction: **s**.

(a)   (b)

Figure 2.19
Same as Figure 2.17a and 2.17b but viewing direction **a**.

(a)   (b)

2.6 DISCUSSION

Scroll waves resulting from unknotted, untwisted ribbons such as that shown in Figure 2.7 have been observed experimentally in the BZ reaction [23]. Scroll waves with unusual shapes have also been reported, but not classified (e.g., in reference 23). We hope to facilitate the interpretation of future experiments with the help of the present work.

An essential outcome of this work is that the technique of cylindrical slicing yields the simplest and easiest-to-interpret graphical representation. It must be kept in mind, however, that this representation is usually not accessible to the experimentalist. On the other hand, the results of planar sectioning may well be compared with experiments such as those obtained with stacks of Millipore filters, soaked with reagent. These filters can be separated and fixed after scroll waves have evolved [21]. An experimental alternative could be to let the reaction proceed in a gel (as in references 29, 52, and 53) and then to cut the gel with a microtome after freezing (or without freezing if the reaction is slow enough). Still another alternative when working with fluorescent reactants would be to shine a quasi two-dimensional beam of light. Planar sectionings recorded by any experimental technique may be used to reconstruct planar or cylindrical slicings with the help of image processing in a fashion similar to that used with microtome cuts in anatomy.

The present work has shown that while planar sectioning renders detailed, localized information, planar slicing gives a more general overview of the topology of the scroll waves. A good example is given in Figure 2.17, in which planar slicing (a) gives a general impression while planar sectioning shows us particular symmetries (b) or asymmetries (c).

The automaton used in this work runs about thirty times faster than the integration of the corresponding partial differential equations [41]. If one considers the capability of this automaton to render quantitative results in agreement with experiments in two dimensions [41,43,44], it is reasonable to propose it as a tool for complex three-dimensional simulations. Interesting and mostly unresolved questions arise regarding the stability of different types of scroll waves. In order to answer these questions, the model should be extended to include excitability of refractory states (as in references 40–44) which was left out here to keep the model simple; calculations of scroll waves including these extensions are underway.

ACKNOWLEDGMENTS

This work was supported by the Commission of the European Communities, Brussels. We thank Gesine Schulte for efficient photographic work and thank Juergen Block for last-minute help with the text processing.

REFERENCES

[1] V.S. Zykov, *Simulations of Wave Processes in Excitable Media*, Manchester University Press (1988).

[2] J.J. Tyson and J.P. Keener, *Physica D* **32**, 327-361 (1988).

[3] A.V. Holden, M. Markus, and H.G. Othmer (eds.), *Nonlinear Wave Processes in Excitable Media*, Plenum Press, London (1991).

[4] A.L. Hodgkin and A.F. Huxley, *J. Physiol.* **117**, 500-544 (1952).

[5] N.A. Gorelova and J. Bures, *J. Neurobiol.* **14**, 353-363 (1983).

[6] K. Hara, P. Tydeman, and M. Kirschner, *Proc. Natl. Acad. Sci. USA* **77**, 462-466 (1980).

[7] V. I. Koroleva and J. Bures, *Brain Res.* **173**, 209-215 (1979).

[8] N. Wiener and A. Rosenbluth, *Arch. Inst.Cardiol. Mex.* **16**, 205-265 (1946).

[9] M.A. Allessie, F.I.M. Bonke, and F.J.G. Schopman, *Circ. Res.* **33**, 54-62 (1973).

[10] M.A. Allessie, F.I.M. Bonke, and F.J.G. Shopman, *Circ. Res.* **39**, 168-177 (1976).

[11] M.A. Allessie, F.I.M. Bonke, and F.J.G. Schopman, *Circ. Res.* **41**, 9-18 (1977).

[12] A.B. Medvinsky, A.V. Panfilov, and A.M. Pertsov, in *Self-Organisation, Autowaves and Structures Far from Equilibrum*, V.I. Krinsky (ed.), Springer-Verlag, Berlin, pp. 195-199 (1984).

[13] G. Gerisch, *Naturwissenschaften* **58**, 430-438 (1971).

[14] A.J. Durston, *J. Theor. Biol.* **42**, 483-504 (1973).

[15] K.J. Tomchik and P.N. Devreotes, *Science* **212**, 443-446 (1981).

[16] F. Siegert and C. Weijer, *J. Cell Sci.* **93**, 325-335 (1989).

[17] A.B. Carey, R.H. Giles Jr., and R.G. Maclean, *Am. J. Trop. Med. Hyg.* **27**, 573-580 (1978).

[18] J.D. Murray, E.A. Stanley, and D.L. Brown, *Proc. Roy. Soc. London B* **229**, 111-150 (1986).

[19] J.D. Murray, *Am. Sci.* **75**, 280-284 (1987).

[20] A.W.M. Dress, M. Gerhardt, and H. Schuster, in *From Chemical to Biological Organization*, M. Markus, S. C. Mueller, and G. Nicolis (eds.), Springer-Verlag, Heidelberg, pp. 134-145 (1988).

[21] A.T. Winfree, *Science* **181**, 937-938 (1973).

[22] K.I. Agladze and V.I. Krinsky, *Nature* **296**, 424-426 (1982).

[23] B.J. Welsh, J. Gomatam, and A.E. Burgess, *Nature* **304**, 611-614 (1983).

[24] B.F. Madore and W. L. Freedman, *Science* **222**, 615-616 (1983).

[25] A.T. Winfree and S.H. Strogatz, *Nature* **311**, 611-615 (1984).

[26] J.P. Keener and J.J. Tyson, *Physica D* **21**, 307-324 (1986)

[27] S.C. Mueller, Th. Plesser and, B. Hess, *Physica D* **24**, 71-86 (1987).

[28] M. Markus, S.C. Mueller, Th. Plesser, and B. Hess, *Biol. Cybern.* **57**, 187-195 (1987).

[29] W. Jahnke, C. Henze, and A.T. Winfree, *Nature* **336**, 662-665 (1988).

[30] P. Foerster, S.C. Mueller, and B. Hess, *Proc. Natl. Acad. Sci. USA* **86**, 6831-6834 (1989).

[31] L.S. Schulman and P.E. Seiden, *Science* **233**, 425-431 (1986).

[32] J. Feitzinger, in *Nonlinear Wave Processes in Excitable Media*, A.V. Holden, M. Markus, and H.G. Othmer (eds.), Plenum Press, London, pp. 351-360 (1991).

[33] U. Frisch, B. Hasslacher, and Y. Pomeau, *Phys. Rev. Lett.* **56**, 1505-1508 (1986).

[34] D. d'Humières, P. Lallemand, and U. Frisch, *Europhys. Lett.* **2**, 291-297 (1986).

[35] J.P. Rivet, M. Hénon, U. Frisch, and D. d'Humieres, *Europhys. Lett.* **7**, 231-236 (1988).

[36] M. Gerhardt, H. Schuster, and J.J. Tyson, *Science* **247**, 1563-1566 (1990).

[37] S. Wolfram,*Theory and Applications of Cellular Automata*, World Scientific Publishing Co., Singapore (1986).

[38] V.S. Zykov and A.S. Mikhailov, *Sov. Phys. Dokl.* **31**, 51-52 (1988).

[39] A.T. Winfree, E.M. Winfree, and H. Seifert, *Physica, D* **17**, 109-115 (1985).

[40] M. Markus and B. Hess, in *Dissipative Structures and Transport Processes and Combustion*, D. Meinkoehn (ed.), Springer-Verlag, Heidelberg, pp. 197-214 (1990).

[41] M. Markus, in *Spiral Symmetry*, I. Hargittai and C.A. Pickover (eds.), World Scientific Publishing Co., Singapore, pp. 165-186 (1992).

[42] M. Markus, *Biomedica Biochimica Acta* **49**, 681-696 (1990).

[43] M. Markus and B. Hess, *Nature* **347**, 56-58 (1990).

[44] M. Markus, M. Krafczyk, and B. Hess, in *Nonlinear Wave Processes in Excitable Media*, A.V. Holden, M. Markus, and H. G. Othmer (eds.), Plenum Press, London, pp. 167-182 (1991).

[45] A.V. Panfilov and A.T. Winfree, *Physica D* **17**, 323-330 (1985).

[46] P.J. Nandapurkar and A.T. Winfree, *Physica D* **29**, 69-83 (1987).

[47] A.V. Panfilov and A.N. Rudenko, *Physica D* **28**, 215-218 (1987).

[48] J. Gomatam and P. Grindrod, *J. Math. Biol.* **25**, 611-622 (1987).

[49] J.P. Keener and J.J. Tyson, *Science* **239**, 1284-1286 (1988).

[50] P.J. Nandapurkar and A.T. Winfree, *Physica D* **35**, 277-288 (1989).

[51] J.P. Keener, *Physica D* **34**, 378-390 (1989).

[52] K.I. Agladze, V.I. Krinsky, A.V. Panfilov, H. Linde, and L. Kuhnert, *Physica D* **39**, 38-42 (1989).

[53] A.M. Pertsov, R.R. Aliev, and V.I. Krinsky, *Nature* **345**, 419-421 (1990).

[54] R.J. Field and R.M. Noyes, *J. Chem. Phys.* **60**, 1877-1884 (1974).

[55] J.-L. Martiel and A. Goldbeter, *Biophys. J.* **52**, 807-828 (1987).

[56] M. Gerhardt and H. Schuster, *Physica D* **36**, 209-221 (1989).

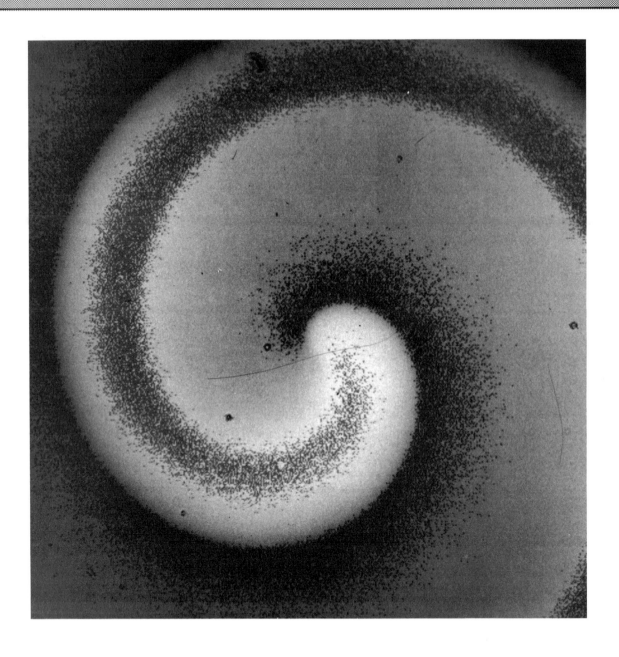

3

Visualization
of Chemical Gradients

Theo Plesser, Wolfgang Kramarczyk, and Stefan C. Müller
Max-Planck-Institut für Moleculare Physiologie
Dortmund, Germany

3.1 INTRODUCTION

Nature displays a large variety of spatial patterns with a wide range of degrees of complexity, from simple and regular to highly disordered forms. Among the fundamental issues of today are the mechanisms of generation and evolution of such structures and their complex functions. Recently, this area of research has been activated by new methodologies and concepts, originating from various disciplines of science [1,2].

One approach to the elucidation of pattern-forming mechanisms is embedded in the concepts and unifying views of nonequilibrium thermodynamics and, more mechanistically, is based on the nonlinear dynamics of chemical and biochemical systems in which the coupling of reaction to transport processes such as diffusion or hydrodynamic convection leads to the formation of spatially organized distributions of the reactive compounds. In particular, reaction-diffusion systems with complex kinetic behavior – for instance, the Belousov-Zhabotinskii (BZ) reaction [3] – have proven to be well suited for the study of patterning processes. In biology, ample evidence is available for application of this concept to regulatory processes at the subcellular and cellular level – for instance, patterns of aggregating cell colonies [4] or cellular differentiation [5].

In the experimental investigation of global chemical dynamics, progress has been achieved in the measurement of spatial patterns – that is, of stationary or propagating chemical concentration gradients [6]. Visualization and quantification of concentration gradients allow for deeper insights into the delicate balance of reaction dynamics and matter transport by diffusion, a matter flow proportional to the negative local concentration gradient. The already mentioned BZ reaction composed of a few simple chemical compounds exhibits fascinating wave phenomena with highly regular geometric structures, like expanding annuli or counter-rotating pairs of spirals. Chemical waves resemble other wave phenomena in nature, such as water waves or electromagnetic waves, with the significant difference that the energy for maintaining the wave motion and its shape (i.e., the gradients) is being provided locally by the free energy of the involved chemical reactions.

An instrument which combines precision optics with video and computer technology (a two-dimensional spectrophotometer) allows the determination of chemical concentration distributions in a solution layer with high spatial resolution. Suitable software routines have been developed to handle the large amount of information provided by this method and to extract geometric and kinetic properties from the digital pattern images [6,7].

In this chapter, we report upon three-dimensional graphic projections for the visualization of two-dimensional data fields and upon algorithms for the calculation of gradients in two-dimensional spectrophotometer data. The techniques visualize structural details and provide new measures for the spatial variation of concentration profiles and gradients.

First, we give a brief description of the instrument and, in some more detail, of the software procedure for three-dimensional graphic representations. It follows an in-depth description of the techniques developed for the proper calculation and visualization of intensity gradients of noisy two-dimensional video images. In the last part of this chapter, examples of patterns and their gradients are presented in a chemical [3,6,7] and a biochemical [8,9] medium.

3.2 TWO-DIMENSIONAL SPECTROPHOTOMETRY

The instrument for space-resolved digital and computerized spectro-photometric measurements consists of ultraviolet-optical components mounted on a vibration-isolated table for illumination and imaging purposes, a video camera serving as the two-dimensional intensity detector, and a fast, large-memory computer for storage of the digitized data and further data processing.

Relevant specifications are (a) the spectral sensitivity of the video camera (Hamamatsu C1000) ranging from 200 to 750 nm, (b) the image raster resolution of 512×512 picture elements, leading to a spatial resolution down to 0.5 μm per pixel at the highest magnification, and (c) the intensity resolution of 256 gray levels per pixel. Fluctuations of the light-sensitive elements of the camera target generate a pixel noise with a standard deviation of ± 3.8 intensity levels. This parameter is independent of the recorded intensity. Acquisition and digital storage are feasible at a frequency up to 30 frames per minute, limited by the time constant of the camera target and the data transfer from the buffer to the magnetic disk. Thus, the apparatus combines spatial, temporal, and intensity resolution in a way which is useful for the analysis of many patterns forming in chemical or biochemical systems. A detailed technical description is given in reference 7.

Each pixel of a frame is identified by its space coordinates x, y – the column and row number – and its intensity $I(x, y)$, an integer number represented by one byte. An experiment with the two-dimensional spectrophotometer results in a time series of typically 50 to 100 video frames occupying 10 to 20 megabytes of disk storage. The efficient analysis of such a large amount of data must rely on a carefully designed software package that combines interactive facilities for visual inspection of the original video data and their various color versions with procedures for automatic processing of long-lasting numerical calculations applied to large numbers of preselected images.

Our software package GRIPS (Graphic Raster Interactive Processing System) provides 80 commands. All of them can be handled on a terminal in an interactive mode or may be edited as a command file for automatic processing [e.g., for unattended (overnight) runs]. The hard- and software system was developed for a Perkin-Elmer 3230 minicomputer and is now in heavy use for research work. Efforts to implement the system on a UNIX workstation are underway.

3.3 VISUALIZATION TECHNIQUES OF TWO-DIMENSIONAL SPECTROPHOTOMETRY DATA

3.3.1 Basic Image Transformations

Large two-dimensional arrays are usually represented as variations in luminosity on a black-and-white or color video monitor [10]. Quantitative information can be visualized by highlighting specific gray levels which convert the image into a black-and-white or colored contour map. A careful selection of pseudo-colors improves significantly the perception of the information contained in the data [11]. Contouring gray levels by colors has the disadvantage that enhancement of one gray level displays the contour as well as other isolated groups of pixels in the image with the same gray level. A flood-fill technique [12] may be useful to highlight only connected contour lines or areas. The following variant of the flood-fill technique has proven to be well suited. The algorithm fills recursively those pixels which fulfill the two two conditions: (A) The gray level is within a specified interval, and (B) the site of the pixels to be filled in the x, y-plane can be reached from a preselected "seed" pixel on the desired contour by a sequence of horizontal, vertical, or diagonal moves touching only those pixels which obey condition A. In this context, "fill" means that the gray level of any pixel identified by conditions A and B is changed to a preset level. Furthermore, analysis by visual inspection or quantitative methods is limited by the pixel noise.

The improvement of the signal-to-noise ratio of an image (taken as a snapshot of a fast-changing dynamic object) is an important operation at the expense of spatial resolution. Most trivial tools are shrinking of the image or, preferably, the calculation of moving averages. Neighborhoods of 3×3 pixels lead to a reduction of the standard deviation of the noise by a factor of 3 (from 3.8 to 1.3 for an uniform image), a usually satisfactory reduction. Averaging by means of much larger pixel neighborhoods would cause an undesirable blurring of the images. More sophisticated smoothing algorithms, such as two-dimensional spline approximations and low-pass spectral filtering in the frequency domain, are discussed in Section 3.3.2 in connection with the computation of derivatives and gradients, the main topic of this chapter.

A more advanced level of visualization is reached if color is combined with perspective three-dimensional presentation [12,13]. The two-dimensional data array of an image recorded by a video cam-

era and stored on the magnetic disk of a computer can be considered
as a bivariate function $I = I(x, y)$ with discrete values given at the
discrete integer coordinates x and y. Our video camera delivers images with $1 \leq x \leq 512$, $1 \leq y \leq 512$ and $0 \leq I \leq 255$. The two-
to three-dimensional transformation allows for looking at the "object"
represented by the data array $I(x, y)$ from different viewpoints as one
would inspect a statue by walking around and changing the altitude
of the eyes. A three-dimensional image makes the gradients of the
function $I(x, y)$ really visible on the two-dimensional display area of
a video monitor. The correct removal of hidden lines and surfaces is
a serious problem in this context, since each pixel of a high-resolution
image has to be checked for visibility. An efficient algorithm [14] has
been developed for this purpose under the following constraints:

1. Orthographic projection

2. Height shading and diffuse illumination so that the intensity of
   any image element remains unaffected if observed from different
   viewpoints

3. Reproduction of the original two-dimensional data field for zero-
   degree rotation and tilt angle

4. Linear interpolation of missing points in the three-dimensional
   projection

5. Achievement of reasonable processing times and elimination of
   the propagation of rounding errors by extensive use of integer
   arithmetic

6. Realization of four types of surfaces: dot images with unconnected
   sample points; profile images with adjacent sample points connected in either the x- or y-direction; crosshatched or grid images
   with adjacent sample points connected in both directions; and
   surface images (i.e., crosshatched images filled by linear interpolation of intensities).

The algorithm traverses the surface from front to rear and thereby
generates pixels along a line in an incremental manner similar to that
of the well-known Bresenham algorithm [15]. Only those pixels are
drawn which passed a visibility test. A pixel is visible if it lies outside
of that area already outlined by the plotted parts of the graph. The
precise definition of that area depends on the type of surface visualization. For dotted and rendered surfaces, the pixel area consists

of gap-free vertical raster-line segments updated immediately after a plot of a new pixel. For profile and crosshatched surfaces the current vertical segment may contain a gap. This gap is removed when the algorithm switches to the next vertical segment. Having recognized this little but essential subtlety, the projected surfaces are correctly displayed on a raster device without drawing any pixel twice.

In addition to the fairly involved algorithm just mentioned, more basic numerical and logic operations are available in the software system. These comprise the fundamental arithmetic operations of addition, subtraction, multiplication, and division which can be performed pixel by pixel for any pair of video frames. Important applications are the additive superposition of constant bias values and multiplicative correction procedures – for instance, the shading correction. This is derived from reference images and accounts for gross imperfections in the optical homogeneity of the illuminating light source and in the sensitivity of the video target. A command for calculating the logarithm of the measured gray levels is frequently used for transforming the spatial intensity distributions of a time series, $I(x, y, t)$, into the corresponding absorbence, $A(x, y, t)$, according to

$$A(x, y, t) = \log[I_o(x, y)/I(x, y, t)] \tag{3.1}$$

The reference values $I_o(x, y)$ are used for the local correction of static spatial distortions in the imaging process.

Equation (3.1) allows the calculation of a spatial concentration pattern according to Lambert-Beer's law

$$A(x, y, t) = \epsilon \cdot d \cdot c(x, y, t) \tag{3.2}$$

where ϵ is the absorption coefficient of the absorbing chemical in the sample, d is the thickness of the sample, and $c(x, y, t)$ is the concentration of the chemical species at the coordinates x, y at the time instant t.

The image transformations are completed by commands for the logical operations "and," "or," "not," "less than," "equal to," and "larger than" for specific evaluation requirements.

3.3.2 Gradient Analysis of Spectrophotometric Video Images

An image recorded by the two-dimensional spectrophotometer can be considered as a discrete representation of a continuous object $O(x, y)$ superimposed with noise:

$$I(x, y) = O(x, y) + N(x, y) \tag{3.3}$$

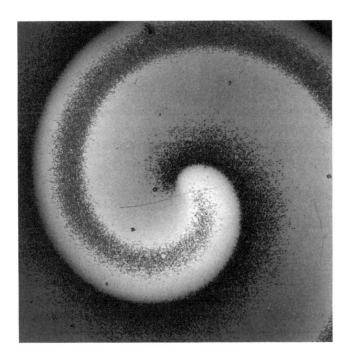

Figure 3.1
High-resolution video image of the center region of a rotating spiral wave in a solution of BZ reaction. The 450 × 450 pixel image shows a 2.1 × 2.1-mm^2 section of the light beam that passed a solution layer of 0.56-mm thickness. The blue color highlights 10% of the total number of pixels counted from the lowest gray level. The red pixels indicate accordingly the 10% portion of highest intensity. See insert for color representation.

The additive noise term N is gaussian in most parts of the image. Only a few disturbances in the specimen, caused for example by gas bubbles or dust particles, may locally generate non-gaussian noise due to over- or underexposure of the camera target.

The data field displayed in Figure 3.1, an image taken from research work on spiral waves in the BZ reaction [16], has been chosen as a realistic model for tests of the various techniques available for data approximation and gradient analysis. A method is required which yields simultaneously a reasonable good approximation and visually appealing graphic representation of the original two-dimensional data and at least first derivatives. This requirement excludes well-known discrete techniques such as Sobel and Laplace filters [13,17,18]. Furthermore, these techniques are known to be highly sensitive to noise.

In the following, two-dimensional splines and fast Fourier transforms will be applied to the test image because these tools seem to be appropriate to reach the goal mentioned in the last paragraph. However, at first the usefulness of spatial averaging within a small pixel neighborhood is shown.

The effect of a moving average with a 3 × 3 pixel neighborhood becomes obvious by a comparison of Figures 3.1 and 3.2. The color blue enhances the 10% of the pixels accumulated starting from the minimum gray level and red enhances the 10% of the pixels counted from the maximum gray level. The color distribution in Figure 3.1 makes the pixel noise really perceptible, whereas the colored 10% levels in Figure 3.2 form connected regions with fuzzy boundaries. Data smoothing by moving average is a fast means to obtain a first impression of structural details of an image by colored contours.

A more general approach is the approximation by means of non-rational B-bisplines – that is, tensor products of B-splines [12,19,20]. A suitable piece of software for our purpose is SMOOPY [21]. The idea behind SMOOPY can be roughly described as follows. SMOOPY seeks the smoothest B-bispline among all those B-bisplines which are close enough to the data. Here, the sum of squares of discontinuities in respective higher derivatives serves as a measure of smoothness of the fit, and the (overall) sum of squared residuals serves as a measure of closeness of the fit. The former sum is minimized, subject to the constraint that the latter sum does not exceed a user-provided value. The algorithm starts with two knots on each coordinate axis and adds successively more and more knots until the overall sum of squares of the residuals reaches the user-specified value. New knots

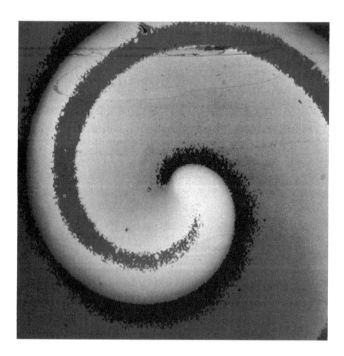

Figure 3.2
The same data as in Figure 3.1 after application of a moving average with a 3 × 3 pixel neighborhood. The colors enhance minimum and maximum intensity as defined in Figure 3.1. See insert for color representation.

are added preferably at intervals with largest column-sums or row-sums of squared residuals, respectively. The closeness-of-fit value allows the user to control the quality of the fit. Too small a value results in an overfit: The B-bispline surface approximates the irregular noise pattern too closely, and as a consequence its derivatives oscillate with high frequencies and amplitudes, leading to sharply wrinkled gradient fields. Too large a value results in an underfit: Fine details are unduly blurred. As suggested in refernce 22, the closeness-of-fit parameter constitutes an estimate of the noise variance. This suggestion leads to a specification of the closeness-of-fit by means of the noise dispersion per pixel, a quantity primarily determined in our case by the pixel noise of the light-sensitive target in the video camera [7,23].

The software, originally coded in FORTRAN 66, has been recoded in well-structured FORTRAN 77 to make it transparent and easy to maintain. The dimensions of SMOOPY's numerous working areas are now specified by very few parameters lumped together in one PARAMETER statement. In addition, more compact data structures are used. The code has been thoroughly optimized to achieve reasonable execution times for images as large as 512×512 pixels. The heuristic knots-adding procedure has been improved. In the original code even a negligible difference between the maximum column-sum and the maximum row-sum of squared residuals could induce the addition of a large number of knots solely in one of the two directions. However, unjustified differences in the number of knots along the two coordinate axes distort the visual appearance of the resulting images. This improved procedure takes care that there is a reasonable balance between the number of knots added for both directions. Finally, a routine was added, which evaluates B-bisplines simultaneously with their derivatives up to a specified order. The algorithm is based on the DeBoor–Cox recurrence relation for B-splines [19,20,24] which guarantees high numerical stability. In summary, the modification and optimization of the SMOOPY software package lead to better two-dimensional spline approximations and to a reduction of the execution times by a factor of up to three compared to the originally published version [21].

Figure 3.3 displays a fit of the data field shown in Figure 3.1 by means of the modified SMOOPY algorithm. The corresponding absolute values of the gradients, calculated from the partial derivatives in x and y direction of the B-bisplines, are presented in Figure 3.4. The bright spiral pattern represents the loci of maximum gradient, whereas the dark contours correspond to the intensity extremes, maxima as well as minima, of Figure 3.3. The spline fit was calculated

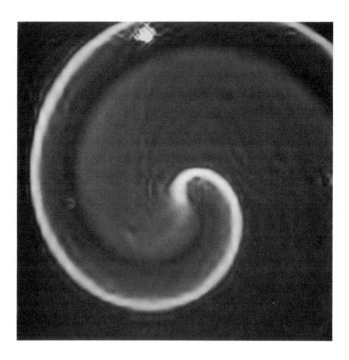

Figure 3.3
Visualization of a B-bispline
approximation of the data
shown in Figure 3.1. The colors
enhance minimum and
maximum intensities as defined
in Figure 3.1. See insert for
color representation.

Figure 3.4
Distribution of the size of the
intensity gradients calculated
from the B-bisplines displayed
in Figure 3.3. Dark contours
reflect extremes (i.e., minima
and maxima in the original
data). Bright contours render
the loci of steepest gradients.

with a noise dispersion of 4.75 gray level pixels, a reasonable value in relation to the experimentally estimated value of 3.8 gray levels [7,23].

It can be seen that some regions are overfitted whereas others are underfitted. Particularly strongly overfitted is a cross-shaped region centered around the largest by far local disturbance, a dust particle in the middle of the upper part of the image. Because that region is particularly difficult to fit, SMOOPY located knots as densely in the one as in the other direction; At those x-values and y-values, over-fit patterns occur even far from the dust particle itself. Thus, in spite of the local nature of B-splines, the effect of a local disturbance is not local in two dimensions, which is a severe drawback.

As an alternative technique, spectral filtering based on the fast Fourier transform (FFT) was tested. Our implementation of two-dimenional FFT follows rather closely the code given in reference 25. The transform is computed almost "in place", that is, frequencies which are redundant due to the hermitian condition are not calculated at all. Thus, the storage requirement is reduced to one-half of the amount needed by most other radix-2 FFT algorithms. Reduction of wrap-around errors was achieved by toroidal image extension to a power-of-2 size and linear interpolation between opposed boundary pixels. More sophisticated interpolation schemes are conceivable.

To obtain smooth images, the high-frequency information, which includes the noise, has to be suppressed. This can be achieved by applying the Butterworth low-pass filter [13]. The filter parameters are tuned interactively. An optimum is reached when the noise-induced lumps just cease to be perceivable in the resulting image. Figure 3.5 displays the approximation of the pattern in Figure 3.1 by spectral filtering. The image is remarkably smooth, and even the dust particle in the upper part is barely visible. The colors highlight the crest and the trough of the spiral and support the impression of a smooth structure.

Partial derivatives are obtained by multiplication of the FFT either by $-2\pi k_x/N_x$ or by $-2\pi k_y/N_y$, respectively, and the subsequent computation of the inverse transforms. N_x and N_y are the number of pixels in the x and y direction and k_x and k_y are the corresponding spatial frequencies. The distribution of the size of the gradients calculated from the partial derivatives are displayed in Figure 3.6. The result is by far superior to that obtained by the spline fit (Figure 3.4).

The execution time of the FFT-based method is about one order of magnitude shorter than that of the B-bispline based method. The computation of the images in Figures 3.5 and 3.6 took about 5 min of CPU time of a minicomputer (Perkin-Elmer 3220).

From the discussion above, it is clear that the FFT-based smooth-

Figure 3.5
Visualization of the spiral in Figure 3.1 after application of an FFT based spectral filter technique. The high frequencies were suppressed with a Butterworth low-pass filter. The colors enhance minimum and maximum intensities as defined in Figure 3.1. See insert for color representation.

Figure 3.6
Distribution of the size of the intensity gradients calculated from the low-pass filter approximation displayed in Figure 3.5. Dark contours reflect extremes (i.e., maxima and minima in the original data). Bright contours render the loci of steepest gradients.

ing technique has many benefits compared to the B-bisplines. Because both techniques process the two-dimensional data field in a different way and different parameters are used to optimize the outcome from the user's point of view, they can complement each other. The application of both methods is strongly recommended if significant scientific evidence is to be based on fine details emerging in the fitted and/or gradient images, respectively.

Finally, the following tools have been incorporated into the GRIPS software mainly as a means for the interpretation of gradient images. The first tool is the interactive flood-fill technique for enhancement of contour lines mentioned above, which is especially helpful to distinguish minima and maxima by different colors. As an example see Figure 3.10 below. The second tool is borrowed from fluid dynamics and implements the visualization of the size and orientation of gradients by arrows. A lattice of scaled arrows overlaid on the original image gives an informative impression of the gradient distribution similar to a two-dimensional flow field. The difficulty of displaying very small arrows on a raster device can be overcome by nonlinear scaling of the arrow size, as suggested in reference 26 and demonstrated in Figure 3.12.

3.4 VISUALIZATION OF GRADIENTS IN REACTION PATTERNS

3.4.1 Spiral Waves in the Belousov- Zhabotinskii Reaction: A Chemical Example

The propagation of chemical waves is observed in the Belousov-Zhabotinskii (BZ) reaction, in which an organic acid is oxidized and decarboxylized by bromate in the presence of a metal ion catalyst (e.g., ferroin or cerium) [3]. In a stirred solution, the reaction oscillates between a reduced and an oxidized state, if the various compounds are mixed in proper proportions. For our purpose, the mixture is prepared in such a manner that a typical 0.6-mm layer in a flat dish remains in a quiescent reduced state but can be easily excited to a local transition to the oxidized state by a perturbation, for instance, by immersing a hot platinum wire or simply by dust particles. The result is an outward propagating circular wavefront of chemical activity, behind which the medium gradually returns to the quiescent state. After a refractory period, this state becomes newly excitable. A disruption of the wavefront by a gentle blast of air ejected from a

Figure 3.7
Symmetric pair of spiral waves in a thin layer of an excitable BZ reaction. The gray levels of transmitted light intensity (490 nm) measured for the 410×410 pixels in panel A are connected by linear interpolation (surface image) and displayed in panel B in three-dimensional perspective at a tilt angle of 45°. The narrow gray level interval enhanced in black is the same in both images. (Reproduced from reference 8)

A

B

micropipette leads with high probability to the formation of a pair of spiral-shaped waves [6,7].

Figure 3.7A shows a pair of such waves with inward-turning tips – that is, with the opposite sense of rotation. A surface image is used for rendering the shape of the intensity distribution in three-dimensional form to get a first impression of the various concentration gradients in the measured data field (Figure 3.7B). In this presentation, one can recognize in detail the height of the wave crest and its modulation at the collision area between both fronts where the two waves start to annihilate each other. In terms of chemistry, the dip at the area of interaction indicates a more reduced state of the catalyst and indicator ferroin produced by the collision. The crest, which marks the spatial transition between oxidation and reduction of the catalyst, reveals local chemical features of the reaction kinetics to be analyzed further.

The intensity level enhanced in black shows that the transformation to three-dimensional perspective retains the isoconcentration lines. It emphasizes the geometric regularity of the pattern which is found to be well described by Archimedian spirals except for the core region close to the spiral tip [6,7].

For further analysis of the innermost area of the spiral, a small section (2.1×2.1 mm^2) containing only the tip and one whorl of one spiral was selected and a time series of 450×450 pixel images covering three rotations was recorded at intervals of 2.8 sec in an experiment. One image from this time series is shown in Figure 3.8A. The same image was used in Figures 3.1 to 3.6 for the demonstration of the various smoothing and gradient techniques. Two iso-intensity lines around a gray level of 175 are enhanced in black. These lines are separated by a gray level interval of 10, while the difference between minimum and maximum intensity of transmitted light amounts to 65 gray levels (compare Figure 3.9). The purpose of enhancing these selected levels is visualized in Figure 3.8B, in which the inner section of image A is digitally expanded. Generally, the distance between the two black iso-intensity curves is small along the steep excited fronts of the wave and larger along the smooth inner slopes of the relaxing back of the wave, as expected. The feature to be emphasized is the remarkable closeness of the lines at a distinctive site near the spiral tip (arrow in Figure 3.8B). This area of very steep gradients is located 54 μm away from, and thus in the immediate vicinity of, the rotation center of the spiral (white circle), which is determined by an overlay of six spiral images [16]. This observation is also supported by another visualization technique (see Figure 3.12).

The intensity profile shown in Figure 3.9A was extracted from

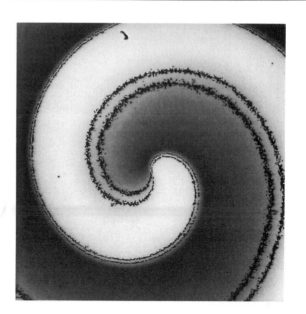

(a)

Figure 3.8
(a) Same data as in Figure 3.1 after application of the moving average with a 3 × 3 neighborhood. Two intensity levels are enhanced in black. (b) Digital expansion of the center section of image A by a factor of 2.5. The white circle indicates the center of spiral revolution. The arrow is referred to in the text. (Reproduced from reference 16)

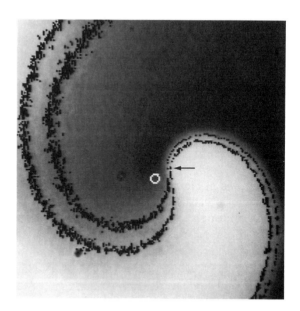

(b)

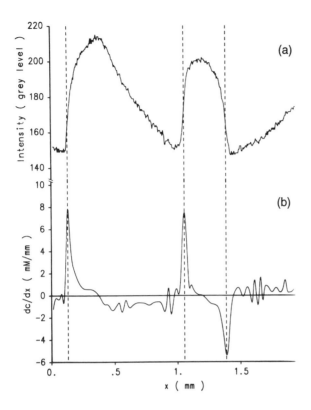

Figure 3.9
(a) Profile of transmitted light intensity extracted from Figure 3.8a and passing horizontally (in x direction) through the site indicated by the arrow in Figure 3.8b. (b) Gradients of concentration of ferriin, derived from profile A by a spline fit and logarithmic conversion to concentration data.
(Reproduced from reference 16.)

the full image of Figure 3.8A and passes in the horizontal direction through this distinctive site. The data points were fitted by splines, and their derivatives in terms of concentration gradients of the oxidized form of the catalyst, ferriin, is plotted in Figure 3.9B. Comparison of the two curves in Figure 3.9 shows that the steep gradients are localized in narrow intervals of the x coordinate. An important observation is that, although the maximum values of about 8 mM/mm are the same for the outward-propagating front (left peak at $x = 1.05$ mm), they occur at different gray levels: 160 for the left peak, which is very close to the intensity minimum between waves, and 175 for the peak in the core region.

Inspection of profiles and their derivatives passing through other points close to the center of rotation reveals that the concentration gradients at the outward-propagating front reach maximum values always at 160 gray levels, whereas the peak values at the inner side of the very spiral tip (i.e., around the site marked by the arrow in Figure 3.8B) occur for higher gray levels, which means for higher oxidation. These higher levels follow a systematic trend within a small gray level interval.

Furthermore, we note the biphasic structure of the wavefront in the region between the zeros at $x = 0.12$ mm and $x = 0.37$ mm. The gradient of the waveback is comparatively low, with an average value of approximately 1 mM/mm.

With the new technique allowing for the calculation and visualization of gradients in two-dimensional perspective, additional features are detectable in BZ waves. In Figure 3.10, the most significant contours of the gradient image shown in Figure 3.6 are marked to facilitate the comparison between the gradients and the structure of the wave pattern. The marked contours are determined by the flood-fill algorithm mentioned above and are colored as follows: The red contour corresponds to the maximum intensity while the small blue line corresponds to the minimum intensity. The cyan-colored contour marks a region in the waveback characterized by a slight increase of an otherwise decreasing gradient. This effect is also visible in Figure 3.6 as a faint contour starting from the tip where the regime of the steepest

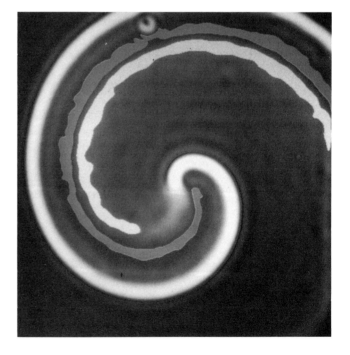

Figure 3.10
Distribution of the size of the intensity gradients. Same picture as Figure 3.6 but with the colored enhancements of significant contours. Red: crest of the wave (intensity maximum). Blue: trough of the wave (intensity minimum). Cyan: special contour on the waveback (see text). See insert for color representation.

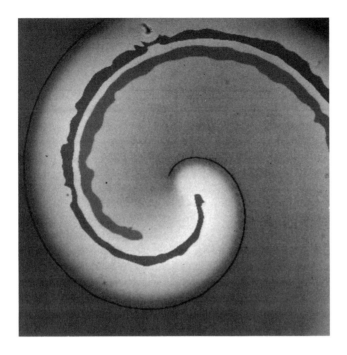

Figure 3.11
Contours highlighted in the gradient image of Figure 3.10 overlaid onto the wave structure in Figure 3.5. Red: crest of the wave (intensity maximum). Blue: trough of the wave (intensity minimum). Cyan: special contour on the waveback (see text). See insert for color representation.

gradients ends. In Figure 3.11, these contours were digitally copied onto the fitted wave pattern (Figure 3.5). In this figure, the color red marks the crest of the spiral, cyan indicates the special region on the waveback, and the thin blue line highlights the boundary between the back of the precursor wave and the leading front of the new wave. In this region of minimum intensity a small gray level interval is enhanced in yellow. It is clearly visible that the distribution of yellow pixels is not homogeneous along the trough; that is, the ferriin concentration varies along the seemingly very regular wave structure. This result supports the interpretation of the dynamics of the wave motion given elsewhere [27].

The visualization of gradients (i.e., their size and orientation) is demonstrated in Figure 3.12. The eddy-like structure in the inner mold of the spiral tip reflects the gradient distribution as discussed above in connection with Figure 3.9. Finally, a three-dimensional perspective view of the intensity distribution of a rotating BZ spiral was computed and presented as a crosshatched surface as shown in Figure 3.13.

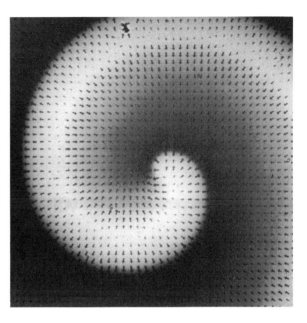

Figure 3.12
Size and orientation of the gradients in a wave structure visualized by arrows. Arrow size is proportional to (gradient size)$^{0.4}$. The white circle marks the rotation center of the spiral.

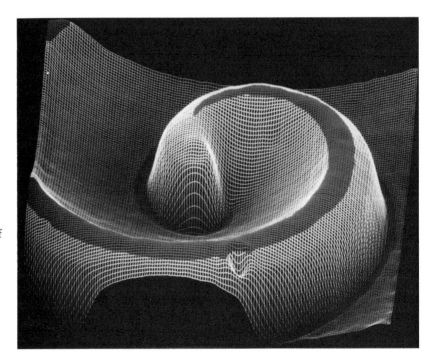

Figure 3.13
Three-dimensional perspective rendering with a grid surface of the spiral computed from filtered data (Figure 3.5). Red: crest of the wave. Blue: trough of the wave. Tilt angle 51°; rotation angle 185° (compare with Figure 3.1). See insert for color representation.

3.4.2 Diffusion in an Enzyme Catalyzed Reaction: A Biochemical Example

Pattern formation by the deposition of an enzyme drop into a layer containing the appropriate substrates is demonstrated with the enzyme lactate dehydrogenase (LDH). This enzyme catalyzes an important reaction step of the anaerobic energy flow in living matter [9]. It oxidizes the reduced form of nicotinamide adenine dinucleotide (NADH) in the presence of pyruvate. If the surface tension of the drop is adjusted to that of the layer (by using suitable buffer solutions), the spreading of oxidation of NADH follows smooth concentration profiles which flatten in time. If the surface tensions differ, then the enzyme is initially deposited in an inhomogeneous manner leading to a more complex pattern of the region of oxidation.

A locally initiated transport process in an enzymatic reaction is demonstrated in Figures 3.14-3.16. A small drop of the enzyme LDH is pipetted into a solution layer containing the substrates NADH and pyruvate at almost equal concentrations. The surface tension of the drop is adjusted to that of the layer, thereby minimizing the disturbances at the moment of drop addition. The two-dimensional image (Figure 3.14) is obtained by NADH absorption as one snapshot from a time series of images. It shows how the initially small, bright center of enzyme activity becomes larger in time while converting NADH to NAD and pyruvate to lactate [16]. The three-dimensional presentation of the filtered data (Figure 3.15) illustrates the regular intensity distribution expected from a diffusion-controlled process. The regularity of the pattern is emphasized by the gradient distribution (Figure 3.16).

Figure 3.14
Diffusion in a 1.8-mm layer of pyruvate/NADH solution after addition of a drop of the enzyme LDH, which catalyzes the biochemical turnover. The image of the light intensity transmitted at 370 nm shows a snapshot taken 8 min after addition of the enzyme.

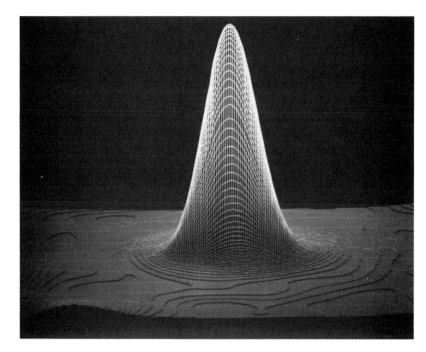

Figure 3.15
Three-dimensional perspective rendering with a crosshatched surface of the intensity distribution shown in Figure 3.14 after spectral filtering. Tilt angle 75°.

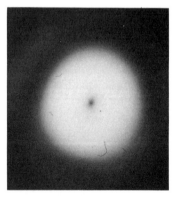

Figure 3.16
Distribution of gradient sizes in the filtered data of Figure 3.14 used for the computation of the three-dimensional intensity representation in Figure 3.15.

3.5   CONCLUDING REMARKS AND OUTLOOK

There is an increasing number of scientific areas in which the phenomenological or quantitative characterization of pattern-forming systems has been recently improved [28]. One important reason for this development is the rapidly increasing efficiency with which video cameras linked to computer systems can be used in the laboratory. This type of digital technique is readily applied to biological research. Examples are the analysis of fluorescence patterns on a cellular scale [29] or improvement of the resolution in classical light microscopy [30]. Quantitative results measured in chemical and biochemical systems have begun to appear in the literature [6].

The latter applications demand new software supporting the extraction of detailed information contained in video images but hidden to some extent in the pixel noise generated by the light sensitivity target of the video camera. Bias-free smoothing techniques are very important for noise reduction before coloring techniques (e.g., pseudo-colors) are applied for the evaluation of fine details. Even more demanding are techniques for the calculation of concentration gradients which, in the context of distributed reaction systems, are a "natural" property governing the matter flow according to Fick's law.

This chapter reviews some of the techniques available in the literature and describes their improvements necessary for a successful application to research work dealing with pattern formation in chemistry and biochemistry. During the preparation of this work, many ideas arose, regarding what could be done to improve our algorithms. One topic is a further reduction of distortions at the image boundaries which occur when applying spectral FFT filters. In addition, new averaging techniques such as binomial filters [17] are under consideration.

The authors believe that, for research focusing on insights into the creation of patterns by nature, the development of powerful analysis and visualization tools is of importance similar to that of the development of experimental equipment for data acquisition.

REFERENCES

[1] P. Glansdorf and I. Prigogine, *Thermodynamic Theory of Structure, Stability and Fluctuations*, Wiley-Interscience, New York (1971).

[2] H. Haken (ed.), Springer Series in Synergetics, Springer, Berlin.

[3] A.N. Zaikin, and A.M. Zhabotinskii, *Nature* **225**, 535 (1970). Also, R.J. Field and M.Burger (eds.) *Oscillations and Traveling Waves in Chemical Systems*, Wiley, New York (1985).

[4] G. Gerisch, *Naturwissenschaften* **58**, 430 (1971).

[5] H. Meinhardt, *Models of Biological Pattern Formation*, Academic Press, London (1982).

[6] S.C. Müller, Th. Plesser, and B. Hess, *Science* **230**, 661 (1985); S.C. Müller, Th. Plesser, and B. Hess, *Physica* **D24**, 71, 87 (1987); C. Vidal, A. Pagola, J.M. Bodet, P. Hanusse, and E. Bastardi, *J. Phys.* **47** 97 (1986); P. Wood and J. Ross, *J. Chem. Phys.* **82**, 1924 (1985).

[7] S.C. Müller, Th. Plesser, and B. Hess, *Naturwissenschaften* **73**, 165 (1986).

[8] S.C. Müller, Th. Plesser, and B. Hess, *Biophys. Chem.* **26**, 357 (1987).

[9] A. Lehninger, *Biochemistry*, North Publishers, New York (1970)

[10] *Raster Graphics Handbook,* Conrac Division, Conrac Corp. Covina, CA 91722 (1980)

[11] P.J. Bouma, in *Physical aspects of colour,* W. de Groot, A.A. Kruithof and J.L. Develtjes (eds.), MacMillan, London (1971); G.M. Murch, *Comput. Graph. Forum* **4**, 127 (1985); G. Wyzecki and W.S. Stills, *Color Science*, Wiley, New York (1982).

[12] J.D. Foley, A. van Dam, S.K. Feiner,. and J.F. Hughes, *Computer Graphics. Principles and Practice,* second edition Addison-Wesley, Reading, MA (1990); D.F. Rogers, *Procedural Elements for Computer Graphics*, McGraw-Hill, New York (1985); J.D. Foley and A. van Dam, *Fundamentals of Interactive Computer Graphics*, Addison-Wesley, London (1983).

[13] R.C. Gonzalez and Wintz, *Digital Image Processing*, second edition, Addision-Wesley, Reading, MA (1987).

[14] W. Kramarczyk, Efficient Raster-Graphing of Bivariate Functions by Incremental Methods in *Theoretical Foundations of Computer Graphics and CAD*, R.A. Earnshaw (ed.), pp. 795-803, NATO ASI Series F, Springer, Berlin (1988).

[15] J.E. Bresenham, *IBM System J.* **4**, 25 (1965).

[16] S.C. Müller, Th. Plesser, and B. Hess, *J. Stat. Phys.* **48**, 991 (1987).

[17] B. Jähne, *Digitale Bildverarbeitung*, Springer, Berlin (1989).

[18] T.S. Huang (ed.), *Picture Processing and Digital Filtering,* Topics in Applied Physics, Vol. 6., Springer, Berlin (1979).

[19] C. DeBoor, *A Practical Guide to Splines,* Applied Mathematical Sciences, Vol. 27, Springer, New York (1978).

[20] M.G. Cox, Practical Spline Approximation, in *Topics in Numerical Analysis,* P.R. Turner (ed.), Lecture Notes in Mathematics, Vol. 965, Springer, Berlin, pp. 79-112 (1982).

[21] Dierckx, *A Fast Algorithm for Smoothing Data on a Rectangular Grid While Using Spline Functions*, Report TW 53, Applied Mathematics and Programming Division, Katholieke Universiteit Leuven, Belgium (July 1980); Dierckx, A Fast Algorithm for Smoothing Data on a Rectangular Grid While Using Spline Functions, *SIAM J. Numer. Anal.* **19**, 1286 (1982).

[22] C.H. Reinsch, Smoothing by Spline Functions (Parts I, II), *Numer. Math.* **10**, pp. 177-183 (1967); *Numer. Math.* **16**, 451 (1971).

[23] S.C. Müller, Th. Plesser, and B. Hess, *Anal. Biochem.* **146**, 125 (1985).

[24] C. DeBoor, Package for Calculating with B-Splines, *SIAM J. Numer. Anal.* **14**, 441 (1977).

[25] W.H. Press, and S.A. Teukolsky, Fourier Transforms of Real Data in Two and Three Dimensions, *Comput. Phys.*, **Sept/Oct.**, 84 (1989); W.H. Press, B. Flannery, S.A. Teukolsky, and W.T. Vetterling, *Numerical Recipes. The Art of Scientific Computing*, Cambridge University Press, Cambridge, England (1986).

[26] H. Duvenbeck and A. Schmidt, Darstellung zwei- und dreidimensionaler Strömungen, in *Visualisierung in Mathematik und Naturwissenschaften*, H. Jürgens and D. Saupe (eds.), Bremer Computergraphik-Tage 1988, Springer, Berlin, pp. 21-38 (1989).

[27] Th. Plesser, S.C. Müller, and B. Hess, *J. Phys. Chem.* **94**, 7501 (1990).

[28] See, for instance: *Science* **221**, 325 (1983); *Electron Microscopy* **226**, 456 (1984); *Paleontology* **226**, 1069 (1984); *Oceanographic observation* **228**, 403 (1985); *Astrophysics* **228**, 597 (1985); *Brain Research* **230**, 663 (1985).

[29] R.V. Tsien and M. Ponie, *Trends Biochem. Sci.* **11**, 450 (1986); N. Stockbridge and W.N. Ross, *Nature* **309**, 266 (1984); J. Lechleiter, S. Girad, E. Peralta, and D. Clapham, *Science*, **252**, 123 (1991).

[30] R.D. Allen, *Annu. Rev. Biophys. Chem.* **14**, 265 (1985).

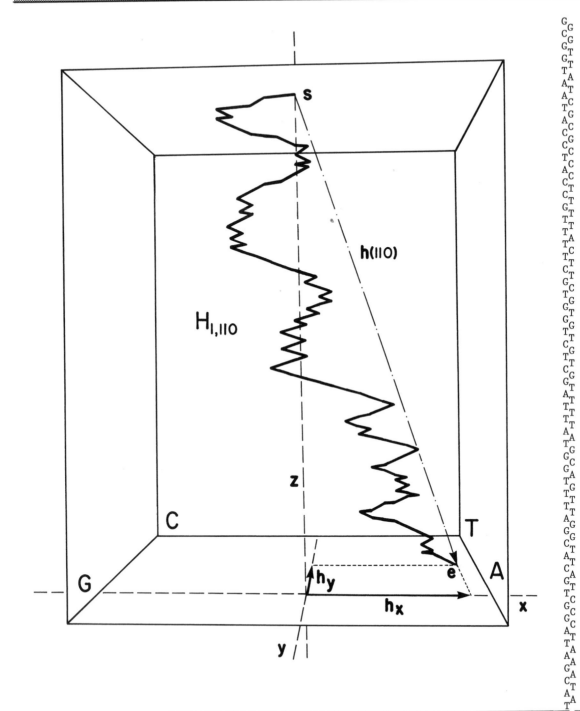

4

Visualization of Biological Information Encoded in DNA

Eugene Hamori
Tulane University Medical Center
Deptartment of Biochemistry
New Orleans, Louisiana

God is in the Details.
Mies van der Rohe

4.1 INTRODUCTION

Deoxyribonucleic acid (DNA) contains the basic genetic information of all living cells. The sequences of bases of DNA (adenine, cytosine, guanine, and thymine – A, C, G, and T) hold information concerning protein synthesis as well as a variety of regulatory signals. The DNA molecule, supported by a twisting sugar-phosphate backbone, contains millions of such bases. In fact, the DNA strings in a single cell would measure up to six feet in length if stretched out, but an elaborate packing scheme coils it to fit within a cell only 1/2500 of an inch across.

The phenomenal capacity of recently developed DNA sequencing methods has opened up the floodgates for the accumulation of biological coding data at an unforeseen pace. Moreover, the output of even the current state-of-the-art sequencing methods will certainly be surpassed in the near future when the worldwide Human Genome Project provides incentives for the development of superfast sequencing techniques.

Our newly found treasure of enormously rich DNA-sequence information is well-defined and extremely precise, albeit monotonous in character. It is commonly represented by a seemingly endless string of letters formed from the variation of four characters, A, C, G, and T, each denoting a nucleotide building block of the DNA. We are far from understanding all the details hidden in this mother lode of genetic information, but we know that the instruction set encoded by the DNA molecules makes possible the carefully controlled functioning of living cells via a sophisticated orchestration of the chemical reactions running within them. Could this massive and complex coding of information within living organisms be directly visualized by human beings, or should it be set aside for endless runs of analyses by computer search programs?

In this chapter, I will discuss certain graphical techniques for finding patterns in genetic sequences. These methods, the G and H curve representations, were developed in my laboratory in order to help the human analyst visualize the information contained in the genetics sequences. As discussed later in this chapter, the esoteric G curves are generated by computers in a virtual five-dimensional (5D) space whose orthogonal coordinates are each assigned to the four DNA nucleotides and to an integer characterizing the position of a nucleotide on the DNA chain. (Because of our inability to comprehend 5D geometry, G curves are useful to us only conceptually and not as a means of visual representations.) A typical G curve is drawn by a computer-graphics program in a virtual 5D space by reading the DNA-sequence data (starting from its 5' end) and directing the G curve along the A axis by one unit if the first nucleotide is A, along the C axis if it is C, and so on. The position axis is incremented by one unit after each nucleotide is drawn. This procedure is continued until the 3' end of the DNA is reached. The resulting line drawing is a continuous curve zig-zagging along the position axis in 5D space. The A,C,G,T coordinate values of the G curve at any chosen point reflect the nucleotide composition of the sequence from its 5' end to the point selected on the DNA sequence.

H curves represent projections of the cryptic G curves into humanly comprehensible three-dimensional (3D) space. They can be visualized as space curves using stereo projections, and they can be rotated, compressed, smoothed, marked, annotated, zoomed to, and so on. The practical utility of H curves is threefold. They can reveal important biological loci, they can be used to study the nucleotide-composition bias of DNA, and they can serve (when furnished with textual annotations) as compact displays of genetic sequence information in a visually comprehensible manner. Unlike the customary DNA letter sequences, H curves do not lose their ability to reflect long-range nucleotide-composition patterns even when reduced to a small size. The application and utility of H curves have been demonstrated on several fully sequenced viral genomes such as the lambda, human immunodeficiency virus (HIV) and Epstein-Barr virus (EBV) genomes in several publications. The method has been also utilized for analysis of large eucaryotic DNA fragments.

The concept of H curves could be developed into a full-fledged graphical metaphor for DNA sequences by taking advantage of the currently available 3D computer graphics technology. The "DNA-conduit" model presented in a subsequent section is intended to be such a metaphor. It would enable researchers to inspect, study, and visually scrutinize vast stretches of conveniently annotated DNA sequences in a familiar 3D environment generated by graphical computer displays.

4.2 NEED FOR A GRAPHIC METAPHOR TO REPRESENT DNA-BORNE GENETIC INFORMATION

Suitable metaphors in the form of graphic images can be very helpful when difficult and complex scientific concepts are confronted. For example, the Feynmann diagrams of particle physics, the conceptual gravity funnels around planets in astrophysics, a chart of the pathways of intermediary metabolism of biochemistry, and the Riemann surfaces of complex analysis are all highly artificial abstract representational devices that are immensely useful to us when we study the behavior of complicated physical systems.

The so-called genetic message borne by DNA molecules in all living cells is one of the fundamental concepts of modern biology. This genetic message (or blueprint) is mainly a set of complex chemical instructions for making proteins and RNA molecules, but the message also contains important regulatory biochemical information. This stored information in a DNA molecule enables it to influence its chemical environment (the cell) such that the reactions in the cell will follow a seemingly purposeful pattern of healthy metabolic activity and not their natural inclination to proceed towards an eventual stagnation in some quasi-equilibrium state. In the peculiar state of metabolites in the cell that we define as living, the chemical environment of the DNA molecules is stable but it is never in equilibrium. For a procaryotic cell, it is the direct consequence of the presence of (DNA-generated) enzymes that the entire chemical system within this simple organism appears to labor towards the re-creation of an appropriately packaged new copy of the original DNA molecule (i.e., the new cell created by cell division).

Our difficulty in comprehending the DNA-borne genetic message lies both with its simplicity and complexity. The stored information is extremely simple because it exists as a linear sequence containing only four coding symbols (A, T, G, and C) designating four nucleotide molecules of relatively simple chemical structure. However, this simple four-letter DNA coding language is rendered extremely complex by the enormous size of the genomic information which typically consists of millions or billions of nucleotides.

What should be the requirements for a good metaphor for this unique DNA-borne genetic message? First of all, the representation should reflect the precise locations of all the functional domains of the DNA [for example: the genes (regions coding for proteins or RNA molecules), including their active and inactive parts (exons and introns); the regulatory regions; the various repetitive DNA regions; domains which are transcribed to RNA by a single operation; and reference points represented by characteristic sites of action of special cleaving enzymes (restriction endonucleases)]. Because an appropriate metaphor should reflect the one-dimensional nature of the stored information, the massive sizes of genomes will certainly present a special challenge. One way to circumvent this problem would be to utilize some zooming feature allowing the representation to be seen in a low-resolution global form, in a high-resolution close-up, and in several intermediate resolution forms.

The DNA sequence carrying the genetic information has an important property which is not apparent from the above discussion. The ratio of the four nucleotides (i.e., local nucleotide composition) in various DNA regions is not statistically balanced near the value of the overall nucleotide composition of the entire DNA. There exist many regions along the DNA which manifest characteristically strong bias in nucleotide ratios. In these regions, one or more nucleotides are present significantly out of proportion (as if along the long DNA sequence there were regional "dialects" of the coding language favoring the preponderance of some nucleotides). Part of this bias may be explained as an exploitation of the flexibility of the 3-letter amino-acid code which permits favoring of certain nucleotides in the genetic blueprint for a particular protein. The biological reasons for the different codon-bias policies along the DNA are not clear, however. They could be manifestations of some control mechanism over the rate of protein synthesis [1 - 3]. The reasons for the imbalance of local nucleotide composition in non-protein-coding regions of the DNA are even less well understood.

The DNA representation metaphor could exploit the regional variation of nucleotide ratios along the entire DNA sequence. Although the meaning of this varying DNA-sequence dialect is, admittedly, not yet understood, the clearly visible nucleotide-composition bias could serve as a conspicuous label for various DNA regions categorized according to the prevailing local dialect (in analogy to the way map makers create geographical regions delineated by various colors assigned according to criteria such as languages spoken, nationalities claimed by inhabitants, diseases prevalent, etc.).

Good metaphors allow intuitive predictions based on the abstract physical image utilized and the DNA-sequence metaphor should have that property also. Finally, the DNA metaphor should be conveniently transportable and communicable among all research-scientist users (unlike, for example, some elaborate mechanical models of physical phenomena enjoyable only by visitors of science museums). In view of these challenging requirements, it is anticipated that a suitable DNA metaphor fulfilling most of the above criteria would be based on advanced computers and would extensively utilize the sophisticated graphics and communication features of these systems.

What is the currently used representation form of DNA sequences against which all new efforts to construct a new and better metaphor must be compared? Molecular biologists, biochemists, and so on, almost always use the standard letter sequence, a seemingly endless string of the four alphabetical characters representing the four nucleotides. For instance, a 600-nucleotide-long part of the human DNA appears (in the format of one of the major DNA-sequence databases, GenBank) as

```
  1 ACCGGCCGGT GTTGTCTGCC ATCTGCAGAC CAGCCCTGCA TAGGCTCAGG ACCAATGACT
 61 GTGGACCTGG GTGTGCATAT GTAGTCCCTG CCACTGTGGT GAATTGCAAA TCAGAGTTTG
121 CAGCTACAGT TGTGTGTTTA GGCTTTGATG CAGGCTGATA CCTCATAATC ACTGAGTTGT
181 TGTTTTCCCA GTTGTACTAT CTTGTGCCTG GACAGTAGCT GTTCTTGGCC TTTTTTCTTT
241 GTGCCTCCTG CTCAGTTACC CCATTAGAGA CTTCGGAGAC TGACCCTGAA TGACTAACTA

301 TTGTCTCCAA GAAGAACTGG AGGCCAATCC ATGACTCTCC GTGGCCATTT TTCTTAAGAC
361 AGAGGCCTGC TTCAATTCTT GACGTATTTA GGGCCCCTGA ATTAAAAACT TGTGCTCTTA
421 CCTGATGTCA AGAAGCACAA AACTCAGATT GCCTCATCCT TTGGACAAGA CCTCTTGGAC
481 TTTGATGGTG TCTCAGGTAC CCTCAACTTT GCTGATCTGG TCAGTTTTCC GTGGTCCCCA
541 CACTAAGAGT CATTCTAACT TGATTGCATC ATGCAATTAT TAGGCTCTTT ATGATATCTG
```

Using that particular character size, a small genome of only 250,000 nucleotides (e.g., that of the recently sequenced cytomegalovirus [4]) would fill about 100 pages of this publication. Obviously, genomes of higher organisms with billions of nucleotides stored on shelves of computer tapes could not be conveniently displayed in print with this format.

One of the major shortcomings of the letter-sequence representation is that global DNA sequence features cannot be visualized. When reading this type of display, one can stand back only to a certain maximum distance from a long letter sequence before the characters become indecipherable. In other words, this representation has only a high resolution format but no intermediate- or low-resolution formats. (One either sees the sequence in its fine details or does not see the information contained in the sequence at all.)

For low-resolution DNA sequence representations, molecular biologists always use some variation of the DNA "line diagram" in which a solid line represents a compressed version of the DNA sequence such as shown in Figure 4.1.

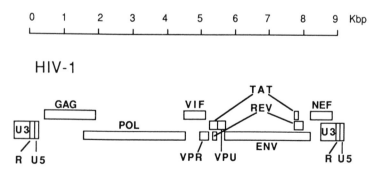

Figure 4.1

An example of a frequently used type of line diagram of a DNA sequence: the genome of HIV-1 (AIDS virus). The rectangular boxes represent identified genes or control elements [5].

These diagrams often come with conveniently annotated markings, arrows, and so on, labeling specific DNA loci. For global representation of sequence features, this primitive metaphor does very well. For DNA molecules which occur in closed loops (i.e., those of viruses and procaryotes), circular line diagrams are particularly useful. Notice, however, that the underlying fine details of DNA sequences are completely missing in these representations. Even for a sequence of only a few thousand nucleotides, one cannot get a magnifying lens, for instance, and scrutinize a particular detail of the DNA sequence. These line diagrams are always fixed at a particular resolution of the sequence information and do not allow zooming in to more details or zooming out to a more global view.[1]

In the following two sections we shall present our work directed towards the development of a suitable DNA-sequence metaphor. Specifically, the basic principles of the methods of G curves and H curves will be outlined. Most details of these studies have been published [6 - 8], including a review with discussion of other graphic DNA-sequence representations [9]. The need to upgrade graphic representations in general, and scientific illustrations in particular, has been eloquently stressed by Tufte [10,11].

4.3 G CURVES: ABSTRACT REPRESENTATION OF DNA SEQUENCES IN FIVE-DIMENSIONAL SPACE

The concept of G curves requiring a geometrical construct in a hypothetical five-dimensional space (at any point of which five mutually perpendicular vectors can be erected!) was conceived with full awareness of the fact that it would be utterly impractical to work with 5D objects even in one's imagination. We also realized, on the other hand, that, by using advanced 3D computer graphics, it would be a relatively easy task to manipulate and work with projections (views) of 5D line constructs in our familiar 3D-space environment.

---

[1]The latter limitation, however, could be resolved if, on a large wall chart for instance, decreasingly smaller and smaller characters would be used for the annotations of local details. These fine details then could be accessed by using some visual aids at close distance. Such line charts, however, are not commonly utilized.

A G curve is an abstract geometrical version of the customary DNA letter sequence. It contains exactly the same information as the DNA letter sequence, but it has several unique properties which are advantageous when representing a lengthy DNA sequence. The G curve is constructed in 5D space by a computer algorithm in the following manner: We assign the four nucleotides A, G, C, and T to four of the five coordinate axes (a,b,c,d) of 5D space and assign the parameter N, the position of a given nucleotide in the DNA sequence, to the fifth coordinate (e). Starting from the 5' end of the DNA sequence and from the origin of the 5D coordinate system, we map the DNA sequence to a 5D space curve as follows: If the first nucleotide is A we move along the a coordinate by one unit, if it is G we move along the b coordinate by one unit, if it is C we move along the c coordinate by one unit, and if it is T we move along the d coordinate by one unit; in order to increment N we also move along the e coordinate by one unit. Thus the coordinates of the first point of the 5D space curve generated will be 1,0,0,0,1 if the first nucleotide is A; 0,1,0,0,1 if the first nucleotide is G, and so on. We proceed, then, to read the second nucleotide of the sequence. Starting from the already established first end point of the curve in 5D space, we arrive at the second point of the curve as follows: If the second nucleotide is A we move along the a coordinate by one unit, if it is G we move along the b coordinate by one unit, and so on; we again move along one unit on the e axis. Accordingly, the coordinates of the second point of the space curve will be 2,0,0,0,2 if the first two nucleotides are AA; 1,1,0,0,2 if the first two nucleotides are AG; 0,0,2,0,2 if they are CC, and so on. This procedure is continued until the 3' end of the DNA sequence is reached. We formally define a *G curve* as the 5D space curve obtained by sequentially connecting all the 5D space points generated in the above manner.

Because of the cumulative effect of steps of the above algorithm, the coordinates of the ith point of the G curve will reflect the nucleotide composition of the sequence from the beginning through the ith nucleotide. For example, for a DNA sequence of 21 nucleotides containing nine A, five G, three C, and four T units, the 5D coordinates of the 21st point of the G curve will be 9, 5, 3, 4, 21. Furthermore, the location of the full set of points (i.e., the shape) of the entire G curve will reflect the precise nucleotide sequence of the whole DNA molecule.

In order to comprehend and visualize this curve of 5D hyperspace, we project it to a specifically oriented manifold of 3D space and designate this familiar 3D image of the original G curve as an *H curve*. Such a 5D → 3D projection can be carried out in an infinite number of ways according to the infinite number of orientations possible of the 3D space within the 5D space. (Using a 3D → 2D analogy, think of the number of ways a cube can be drawn on paper.) Of these projections, however, only three will be directly meaningful and easily comprehensible to most of us. These three are designated as

C T	G C	T G
G A	T A	C A

(The meaning of such groupings of the letters AGCT will be explained below.) In each of these three cases the 3D space receiving the projection is oriented in such a manner that its xz and yz planes are aligned respectively with the two pairs formed from the a, b, c, d coordinate axes of the 5D (donor) space. (In the protocol followed, the e axis of the 5D space is always projected isometrically to the z axis of our 3D space.) For example, the 5D → 3D projection brought about by aligning the xz plane of 3D space with the plane of the 5D space defined by its c and a axes, as well as by aligning the yz plane of the 3D space with the bd 5D plane, will result in the first type of projection.[2]

Most of the relevant properties of G curves can be studied through their 3D H curve projections. The latter, therefore, should be considered as both accurate and humanly comprehensible views of the original G curves.[3]

In the following section we will discuss the properties of H curves on their own without making further references to their descent from hard-to-grasp G curves. It is to be noted, however, that it would be feasible to inspect a 5D G curve directly on a monitor screen by some advanced computer-graphics programs. In the same manner as a 4D hypercube can be studied and scrutinized via its instantaneous 3D projections by a special sophisticated graphics software (e.g., the "Hypercube" demonstration program of Stardent, Inc.), a computer program could be written that would enable the user to view directly on the screen any on-line-generated 3D projections of a G curve using stereoscopic visualization techniques. Because most of these images (excepting the three special views shown above) would be very diffi-

---

[2] Strictly speaking, each of the three types of projections listed above also has a mirror-image twin, but we are not considering these duplicates as inherently different, and therefore separate, projections.

[3] Keep in mind that we very frequently resort to an exactly analogous indirect approach when visualizing 3D objects; we scrutinize them in various 2D projections on paper or on computer screens.

cult to interpret, there is currently no compelling reason to construct such a complex graphics computer program.

4.4 H CURVES: HUMANLY COMPREHENSIBLE VIEWS OF G CURVES IN OUR FAMILIAR THREE-DIMENSIONAL SPACE

Formally, H curves could be considered as convenient 3D projections of the hard-to-visualize 5D G curves which were discussed in detail in the previous section. However, H curves can also be derived directly from DNA sequences without resort to any spatial geometry higher then 3D. (In the latter approach, however, there is a semblance of arbitrariness in the specific assignment of bases to four directions in the 3D coordinate system, and, furthermore, the occurrence of the three varieties of H curves mentioned above is left unexplained.) The direct derivation of H curves is rather straightforward, and it will be presented in the following. It should be noted again, emphatically, that the appreciation of the utility of H curves as compact representations of long DNA sequences is not dependent on understanding their descendancy from G curves.

An H curve is a continuous 3D space curve whose elementary units are the four *base vectors*, each representing one of the four DNA nucleotides A, T, G, and C (Figure 4.2). During the construction of an H curve, these base vectors are joined head-to-tail according to the nucleotide sequence of the DNA. The process starts at the 5' end of the DNA sequence (top of the H curve), and the curve descends towards the 3' end of the sequence. Normally the direction of base vectors used are those shown in Figure 4.2 (standard assignment), but two other assignments are also possible (i.e., the three different groupings of four letters AGCT mentioned above which correspond to the top view of letter positions shown in Figure 4.2). As was discussed above, the different choices of base-vector arrangements correspond to the three inherently different orthogonal projections of an ancestral 5D G curve into 3D space.

An H curve is normally displayed within a rectangular 3D reference box and, depending on the resolution (i.e., number of nucleotides drawn per vertical distance), it typically appears as either a sharply zig-zagging or a relatively smooth space curve descending from top to bottom (Figure 4.3). It is an important fundamental property of H

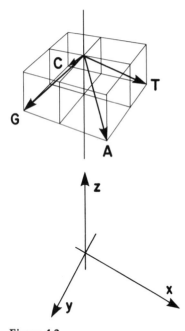

Figure 4.2
One of the three selectable base vector assignments of the four DNA nucleotides A, G, C, and T used in H curves: the standard assignment. (From reference 8.)

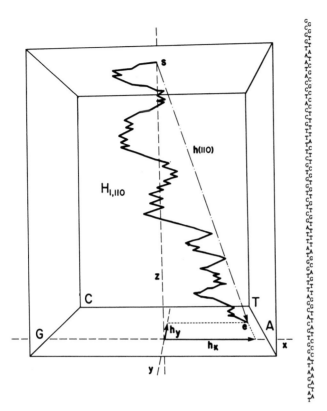

Figure 4.3

Perspective front view of a short H curve representing the nucleotide sequence shown vertically on the right. The total number of nucleotides is 100. The vector $h(110)$ originates at the 5′ end (s) and terminates at the 3′ end (e) of the DNA fragment. The vectors h_x and h_y shown on the x-y plane are the x and y components of h. The capital letters at the bottom corners of the reference box indicate the directions of the base vectors. (From reference 8.)

curves that regions pointing toward a particular vertical edge of the reference box indicate that the region concerned is rich in a specific nucleotide. For example, a curve fragment pointing toward the T edge of the reference box would indicate a preponderance of T nucleotides in that fragment. The 3D nature of the curve can be conveyed to the observer either via two stereoscopic images or by dual presentations of two orthogonal views such as front and side elevations. The choice of the orthogonal viewpoint (or, equivalently, a 90° rotation of the curve and box) will impart special significance to the observed drifting of curve regions towards the right or left sides of the reference box. For example, in the standard front view, regions drifting right indicate AT-rich nucleotide compositions whereas those moving to the left indicate GC-rich ones. Similarly in a side view (from right), drifting of the curve right or left respectively indicates pyrimidine or purine dominances in the nucleotide composition of the fragment. Note that the alternate assignments of base-vector directions mentioned above allow a different partitioning of pairs of nucleotides to the right or to the left box sides (e.g., GT/AC, TC/GA, etc.).

In addition to straight front or side views, other viewing directions can also serve useful purposes. For example, the straight-down top view of the box and its H curve will offer a uniquely condensed view of the sequence that can serve as a characteristic fingerprint of the DNA region involved. Oblique views from various directions are often useful for the scrutiny of some special parts of the H curve with characteristic curvatures. H curves have interesting symmetry properties, and consequently some larger palindromic sequences can be visually identified on them. H curves can be presented in smoothed versions which deemphasize local nucleotide variations. They conspicuously display repeating sequences [8]. However, the most important aspect of H curve presentations is that even very large DNA sequences can be represented within a small space. Despite the fact that in such low-resolution presentations, most of the local details are lost to the unaided eye, the *global* characteristics of the large sequence (typical changes in nucleotide-composition bias, the conspicuous pattern of long repeating sequences, etc.) remain clearly visible to the naked eye.

H curves are constructed, displayed, and manipulated by graphics programs running on computers [12]. In a typical operation, the nucleotide sequences used for H curve generation would be downloaded from a nucleic-acid database via a computer network. Following the calculations, the H curve would be quickly plotted on the monitor screen and the user would be offered a set of options (curve rotations, markings of certain nucleotides, alteration of the smoothing and expansion parameters, output redirection to the XY plotter, etc.) before the second plotting.

There also exist auxiliary programs that will display in rapid succession a previously captured series of incrementally rotated images on the computer screen. In this manner a slowly rotating H-curve/reference-box display is generated which conveys an excellent 3D image. The technical details of the H-curve-generating software used (HYLAS)[4] have been recently described [12].

---

[4] There exist Unix, IBM-VMS, and DOS compiled versions of the original FORTRAN program.

4.5   H CURVE APPLICATIONS

4.5.1   Display of Biologically Relevant Sequence Information

The following research applications cited from studies completed in the author's laboratory are illustrations of the utility of the H curve method.

Indication of Changes in the DNA Template-strand Transcribed

One of the first fully sequenced genomes was that of the lambda bacteriophage [13]. The H curve of this 48,502-nucleotide-long sequence is shown in Figure 4.4. The positions marked by arrows coincide with loci on the genome where the template is switched from one DNA strand to the other [14]. On the side view of the H curve (not shown here) these locations also stand out conspicuously. The occurrence of such prominent breaks in the H curves can be explained by the apparent propensity of this particular phage to create, from either DNA strand, template RNA transcripts which have very similar nucleotide-composition biases [6].

Overlapping Genes

In those rarely occurring short DNA sequences which simultaneously code for *two* different protein fragments, the choices available for synonymous codon selections should be more constrained than the corresponding choices available in non-overlapping genes. This imposed limitation should be manifested in the restriction of the prevailing codon bias and, in turn, the alteration of the current nucleotide-composition bias in the overlapping region. In the short regions of the overlapping genes, the prevailing quasi-uniform nucleotide-composition distribution (if such uniformity exists) should be interrupted by anomalies. Because the spatial direction of H curve regions is an accurate measure of the local nucleotide composition, H curves would be expected to reflect these anomalies in a conspicuous manner. We have done a preliminary H curve analysis of several overlapping genes and our findings support the above expectations [15]. Figure 4.5 illustrates the conspicuous appearance of overlapping gene regions on two H curves (see caption for further details).

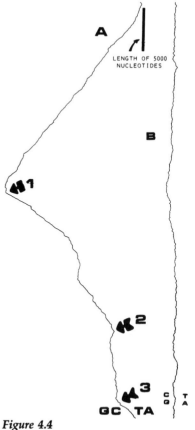

Figure 4.4
(a) H curve of the bacteriophage Lambda genome (front view). (b) The H curve of a hypothetical DNA of the same length and nearly identical nucleotide composition as the Lambda genome, but with a random distribution of nucleotides. (From reference 6).

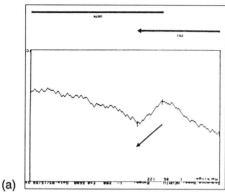

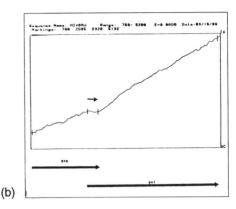

(a)　　　(b)

Figure 4.5
(a) H curve of the overlapping CO1/URFN region of the *Neurospora crassa* mitochondrial genome. The 5′ end of the 280 nucleotide sequence shown is on the left. Rising curve fragments indicate purine-rich regions. The length of the overlap marked by the arrow is 37 nucleotides. (b) H curve of the gag/pol overlap region of the HIV-1 genome, isolate BRU. Rising curve fragments indicate AT-rich regions on the H curve of the 4450-nucleotide-long DNA section shown. The overlap marked by the short arrow extends over 243 nucleotides. (From references 15 and 9).

Repetitive Sequences

In eucaryotic DNA, the presence of overwhelming amounts of non-coding sequences such as introns, satellite and interspersed repeats, and so on, makes the functional characterization of long-sequence domains very complicated. Often, however, useful clues to gene organization can be obtained from the inspection of the H curves of these long DNA sequences. H curves were found to be very sensitive visual indicators of longer sequence repeats [8,16]. Figure 4.6 illustrates the identification of a group of so-called *Alu* sequences embedded in an intron zone of a human DNA sequence. The fingerprint of the ubiquitous 300-nucleotide-long *Alu* sequences is revealing, in spite of the fact that some of them are inserted in reverse orientation and in truncated forms [17]. (*Alu* fragments belong to the family of interspersed DNA repeats. They occur in the human genome more than a half million times and appear to be transposable genetic elements capable of originating DNA replication.)

Global Comparisons Among Sequences

The DNA sequence details of each genome contain the genetic blueprint of the organism. However, there is a hidden *global* characteristic of this information which is not directly evident from the visual inspection of most computer-based analyses of the DNA letter sequences. The changing pattern of nucleotide-composition distribution along the genome is often a unique characteristic of genome families which is always reflected in their respective H curves. Even though the biological significance of most of these patterns is not yet understood, they

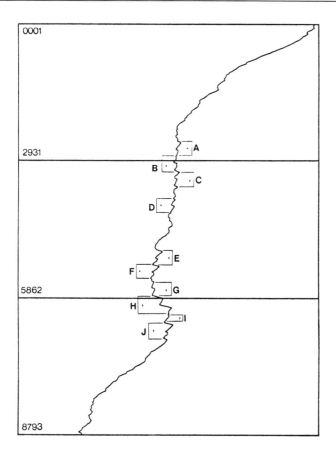

Figure 4.6
The H curve of an
8793-nucleotide-long human
DNA sequence (an expressed
isotype of the beta-tubulin
cluster). Front view with GC to
the left and AT to the right. The
central vertical portion of the
curve corresponds to an
unusual intron cluster of ten
300-nucleotide-long *Alu*
sequences (sections marked
with letters). The horizontal
reference lines are 2931
nucleotides apart. The curve
was smoothed with a *w* factor
of 10. (From reference 7).

can be very useful in sequence comparisons. Figures 4.7 and 4.8 show
the H curves of several HIV genomes and other retroviruses [18]. The
illustrated degree of similarity or dissimilarity among the sequences
selected is very informative. Note, for example, in Figure 4.7 the similarity between the gross features of the *gag* regions versus the marked
dissimilarity of the *gag/pol* overlap regions of the Visna virus and the
HIV-1 genomes, respectively. The analysis of the S3 region of the
EIAV H curve and a similar (non-AT-rich) region of the HIV-1 curves
(open arrow on curve 3 of Figure 4.7), which are both equidistant from
their respective *gag*-start positions, revealed completely different coding functions. This observation led to the suggestion that both regions
might have roles in the packing of the genomic RNA molecule in its
capsid [18].

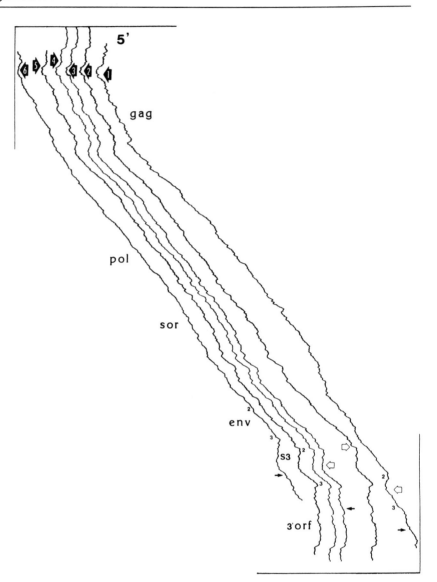

Figure 4.7
The comparison of low-resolution H curves of several HIV and related viral genomes. AT right, GC left (front view). The horizontal positioning of the curves is arbitrary. Curve 1, Visna virus sequence; curve 2, HIV-1 SF2; curve 3, HIV-1 of GenBank locus HIVPV22; curve 4, HIV-1 of GenBank HIVBH101-2; curve 5, HIV-1 of HIVBRU; and curve 6, the equine infectious anemia virus sequence. The small numbers refer to so-called hydrophobic elements in the sequences, and the small solid arrows mark the 3′ ends of the *env* genes. (From reference 18 where further details can be found.)

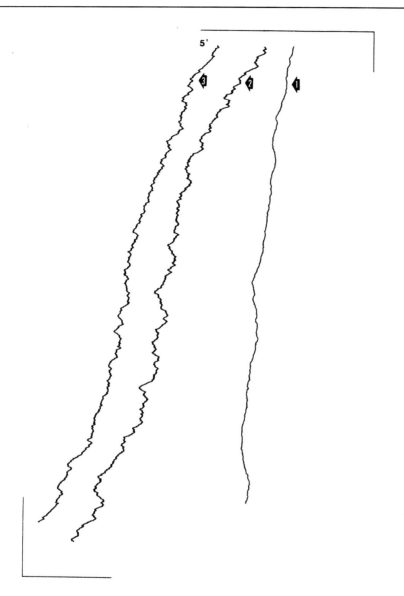

Figure 4.8
Low-resolution H curves of three other retroviral sequences not directly related to HIV-1. Same views and linear nucleotide densities as those of Figure 4.7. Curve 1, Mo-MLV; curve 2, HTLV-1; and curve 3, BLV. (From reference 18 where further details can be found.)

4.5.2 Special Capabilities of the H Curve Method

Compression of Sequence Information

The one-dimensional character of the enormous amount of information carried by DNA molecules has necessitated the evolution of complicated compaction schemes which are manifested in most cells. For example, in the human chromosome the DNA double helix, which is only 2 nm thick but about 10 cm in length, is elaborately folded through five or six levels of microstructural organization into a microscopic volume which is only a small fraction of the approximately 10^{-10}-ml total volume of the nucleus. It is an indirect but interesting parallel that H curves are also able to condense the inherently one-dimensional DNA-*sequence* information to a very compact but still highly informative representational form.

Unlike the traditional letter-sequence representation which is either fully readable or entirely unreadable to the naked eye, H curves can be condensed to a very great extent without losing their characteristic global features. In order to study this particular property of H curves, a few years ago we generated a single continuous H curve of the entire 172,282-nucleotide-long EBV genome by manually assembling several large H curve fragments that we could manipulate in our computer at that time [16]. We found that, in spite of the drastic size reduction used, the characteristic global nucleotide-distribution features of the genome were still recognizable and several loci of direct biological importance [19] could be identified.

We recently generated on the display monitors of both our Sun 386i workstation and our IBM-AT microcomputer the full H curve of the cytomegalovirus DNA sequence of 229,354 nucleotides, which is the current record for fully sequenced genomes [4]. The calculations were done remotely on the Cray Y-MP supercomputer of the Pittsburgh Supercomputing Center, and the graphical output was directly transmitted in its entirety through the Internet computer network. The hard copy of the H curve shown in Figure 4.9 was obtained by capturing the graphics screen output and printing it on a laser printer equipped with a PostScript option. Similarly to the EBV genome, this massive H curve also shows characteristic landmarks at several biologically important loci. The cytomegalovirus genome has a nearly even ratio of purines (G,A) and pyrimidines (C,T), but its composition is strongly biased for the G,C as opposed to A,T nucleotides. These characteristics are clearly reflected in the 3D shape of the H curve, which runs confined to a plane parallel to the front face of the guide box but drifts very strongly to the left-hand side.

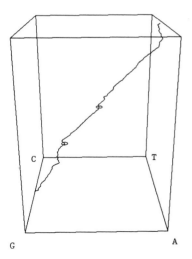 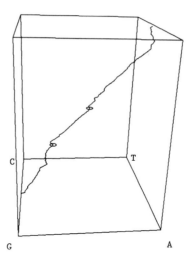

Figure 4.9
Stereoscopic diagram of the H curve of the 229,354-nucleotide-long cytomegalovirus genome. The two horizontal circles mark the locations of the gH transmembrane glycoprotein [20] (upper circle) and the ORF in the transforming domain mtrII [21] (lower circle). The vertical length of the solid rectangle in the upper left-hand corner is equivalent to 7000 nucleotides on the H curve.

Annotations

Published DNA sequences are sometimes embellished with extensive annotations designating the sites of known biological functions, genes, exons, introns, and so on. Our 3D H curves can also be annotated in a similar fashion. They can be marked at various nucleotide locations by small circles drawn in the plane perpendicular to the direction of increasing nucleotide numbers. Furthermore, for computer monitors, printers, or plotters with color capabilities, various H curve sections can be marked in different colors. Future H curve-generating software will certainly take advantage of the sophisticated graphics programs now widely available to include textual annotations and markings. Such annotations would be directly associated with the H curves presented on the computer screen or plotter. The advantages of H curves furnished with state-of-the-art graphic annotations would be particularly apparent for the major DNA sequence projects which have already been initiated (e.g., the human genome project).

Ease of Manipulation

The numerous possible orientations of an H curve allow the inspection of the sequence information from various points of view. As was pointed out above, in one particular view of a 3D H curve, the right/left drift of the curve can mean purine/pyrimidine dominance changes, while in another view the right/left drift means AT/GC dominance changes.

On the H curves of long DNA sequences the global features are always visible but not the fine details (e.g., patterns of small oligonucleotides, palindromic sequences, etc.). However, it is relatively easy for the computer display to zoom into the desired location of such fine sequence detail. (Even single base vectors representing individual nucleotides can be made visible in this manner – for example, Figure 4.5A.) Future versions of H curve generating programs should also have the capability of comparing several different DNA sequences side-by-side and automatically rotating and/or translating the curves in a parallel fashion for optimal views during the analysis (e.g., Figure 4.7).

Display of Codon Bias

Codon bias (also referred to as *nonrandom codon usage*) is manifested over various regions of the DNA as a persistent and almost constant bias of the regional nucleotide composition favoring a particular unbalanced ratio of the four nucleotides. This skewing of global DNA composition is superimposed over the local changes of nucleotide composition (the latter typically originating from the statistical variation of amino-acid code words). Although the biological significance of codon bias is still not fully understood [22], it is believed to be related to the regulation of protein synthesis and gene expression through specific targeting of various groups of tRNA molecules [1-3]. Within a lengthy protein-coding region of the DNA, the direction of the H curve in 3D space is more likely to be a reflection of the codon bias prevailing in that particular region of the DNA molecule than the manifestation of any other effects. (In eucaryotic DNA the number of sequences involved in protein or RNA coding is overwhelmed by the excess of noncoding or defunct coding regions. Typically, these noncoding regions also have biased nucleotide compositions whose significance is even less clear than those of the coding regions.) In spite of the fact that many computer algorithms can also evaluate nonrandom codon usage and predict (with some degree of uncertainty) the presence or absence of protein coding regions [22-24], the *direct* global visualization of nucleotide-composition bias by H curves appears to be of considerable utility. The benefits of the H curve approach will increase even further in the future when the fundamental principles that govern nonrandom codon usage are understood in more detail.

The zones of skewed nucleotide compositions occurring at various locations along a DNA sequence are directly responsible for the characteristic shapes of very large H curves (e.g., Figure 4.9 here and Figure

10 in reference 9). Thus, H curves displaying diverse nucleotide-bias patterns should also serve as characteristic fingerprints for such large sequences.

4.6 OUTLINE FOR AN H CURVE-BASED DNA-SEQUENCE METAPHOR: THE DNA CONDUIT

Can H curves be the starting point for a working metaphor for long DNA sequences? If so, how should their physical presentation be optimized for this particular goal? In this section a plan will be outlined which envisions a "walk-through" H curve model for a very large DNA sequence. In this particular design, the long H curve is imagined to be encased in a rectangular duct (conduit) facilitating the global survey of very large sections of the DNA, and the actual DNA sequence is represented as a thin H curve running at the center of the duct. (The terms *duct* and *conduit* will be used interchangeably in the following discussions.) The model was conceived for a genome of 100 million base pairs, a suitable intermediate size between procaryotic genomes and those of the higher eucaryotic organisms.[5]

The generation and manipulation of the virtual 3D model envisioned here is well within the reach of current 3D computer-graphics technology. The model is designed for at least two distinct and separate inspection modes, namely, a global mode and a high-resolution mode. In the global mode, the visualized image could encompass a very large DNA domain ranging from about 15,000 nucleotides up to the entire genome of 100 million nucleotides. In the high-resolution mode, the fine details of the DNA sequence down to individual nucleotides would be made visible. (Of course, the global aspects of the sequence would not be directly accessible for inspection in this particular mode.) The entire model is based on an H curve drawn in space sideways with the the z axis (representing the number of nucleotides) extending from left to right.

4.6.1 The Global Mode of Representation

The global mode of representation would be accomplished by enclosing the thin H curve in a ductlike conduit of square-shaped cross

---

[5] For example: the *E. coli* bacterial genome has 4.7 million base pairs, the smallest human chromosome has about 16 million, and the entire human genome has about 3 billion base pairs.

Figure 4.10
An H curve based graphical metaphor for long DNA sequences: thin wire running inside a large rectangular duct. The curving of the duct reflects long-range variations of the nucleotide composition of the DNA sequence. In the alternate, high-resolution presentation mode (not shown) the H curve would be greatly magnified and even individual base vectors could be made visible.

section. The depth and height of the conduit (its short dimensions) would be aligned with the x and y axes of the H curve contained within (e.g., AT up, CG down, and GA towards the viewer – when using the standard base-vector directions). Its length would extend horizontally, following the progress of a smoothed equivalent of the original H curve as it propagates in the general direction of the z axis (Figure 4.10). The extent of H curve smoothing [8] that would modulate the short-range direction changes of the conduit would be such as to assure that the H curve inside the conduit would not touch the inner surface of the conduit even at the sites of some strong local nucleotide-composition bias.

For the purpose of presenting this walk-through model, it will be helpful to assign some real scale both to the conduit and to the H curve contained within. Furthermore, the model should be envisioned as a tangible object built from appropriate materials even though its actual realization would be only in the form of computer-generated virtual objects. The length of the H curve unit vectors would be set to 0.75 mm (the size of a spark-plug gap). The H curve should be imagined to be made out of a very thin wire of 0.1-mm diameter. The size of the rectangular conduit surrounding the thin H curve wire would be 2m × 2m (corresponding to a 20,000-fold difference in scale between the thickness of H curve wire and that of the duct). In the global-mode presentation of the model, only the conduit would be made visible by the graphic computer system.

The length of the conduit for the 10^8 bp genome under consideration would be 75 km (47 miles). Thus the scrutiny of the large-scale nucleotide-composition variations of this genome would be equivalent to inspecting the shape of the above conduit suspended high up in the air between Washington, D.C., and Baltimore, MD. The shape of the duct would accurately reflect the large-scale variations in the nucleotide composition of the entire DNA sequence. The locations of some conspicuous direction changes on the conduit would also serve as convenient landmarks for navigation along the lengthy sequence. The primary purpose of the duct device would be to mirror exactly such global DNA-sequence characteristics. Another useful purpose, however, would be the facilitation of sequence annotations. We can envision, for instance, textual annotations on this large-scale model as large billboards posted on the duct designating major biological reference points on the genome (approximate positions of selected gene markers, starts and ends of major transcription regions, locations of extensive repetitive sections, etc.). For the observer approaching the conduit from a distance, these big billboards describing gross sequence features would be noticeable first. But as the observer gradually approached, medium-size signs would also become readable. These would designate somewhat smaller local regions of biological interest (starts and ends of various genes, locations of introns, restriction-endonuclease cleaving sites, etc.). Finally, upon further approach, the observer would clearly discern the actual rectangular shape of the conduit and would reach the limit of the global presentation mode. At that point, the alternate high-resolution mode of the display would have to be activated for the scrutiny of short-range details. This would be equivalent to physically getting inside the duct and directly inspecting the details of the thin H curve wire running inside.

4.6.2   High-Resolution Mode

Because of the illusion provided by the computer-graphics program, one would feel as if he or she were inside the 2-m square rectangular conduit and could directly observe the H curve from a close distance as a brightly illuminated thin wire. Again the tortuosity of the line would reflect the nucleotide-composition variations along the DNA sequence, but in a scale much smaller than that during the outside view of the conduit. Inside the duct, for instance, a 1-m length of the H curve would contain only about 1300 nucleotides. The 229,354-bp cytomegalovirus genome (Figure 4.9), currently the longest completely sequenced viral DNA sequence, would span only a 172-m section of the conduit (about twice the length of a football field). Color coding of the H curve wire would conspicuously mark various sections of the DNA, and the observer would also have the choice of highlighting different regions of special interest. In addition, the thin H curve line could be marked at important locations by various tags (numbers, letters, circles, arrows, marks, etc.) which could be easily created by a 3D graphics program based on the new PHIGS standard. At this resolution, the 0.75-mm-long base vectors designating individual nucleotides would be clearly discernible to the naked eye. They would appear to be very small, however. A further and final enhancement of the high-resolution mode could be initiated by touching (i.e., mouse clicking) a desired locus on the H curve. This signal would activate a slightly enhanced version of the high-resolution display in the form of an annotated letter-sequence projection on the inside wall of the conduit. Projecting the letters at 1.5-mm size (17 pitch), the magnification realized by this enhancement would be twofold relative to the H curve. In this full expansion of the DNA-sequence detail, a 1-kb-long sequence would be 1.5 m in length on the scale of the model.

Partitioning the original model into high- and low-resolution modes would be advantageous from the point of view of graphics computing. This approach would allow a decrease in the number of graphics primitives required to draw views of the H curve conduit from various directions, and this, in turn, would significantly lower hardware requirements.

The variability of base-vector assignments (see previous section) could also be implemented in the H-curve-in-the-conduit metaphor. In the global resolution mode, changing the base-vector assignments would result in an immediate reshaping of the entire duct. The orthogonal axes of the square cross section of the conduit would have new meanings (e.g., from AT excess up, GA excess towards the viewer to TC excess up and GC excess towards the viewer). An analogous situation would prevail when changing the base-vector assignments in the high-resolution mode (inside the conduit). Such switching of base-vector assignments resulting in changes in H curve shapes would not affect the ancestral G curve (hiding in the background, as it were). The new shapes we would observe emerging would be merely the consequences of generating different 3D projections of the original 5D G curve for our convenience.

4.7   H CURVE BUNDLES: GRAPHIC IMAGES OF BIOLOGICAL FUNCTION

In all parts of this section, the entire topic of graphical DNA-sequence representations has been approached strictly from the point of view of DNA sequences. It was shown how G curves and, particularly, H curves could be utilized to represent DNA sequences in a compact and visually surveyable manner. Also discussed was the biological meaning of the shapes of the curves thus generated. However, the same issue can be approached from the opposite point of view as well: How would a set of genetic instructions describing a certain biological function look in the world of DNA sequences? This change in perspective is by no means trivial because the link between biological functionality and DNA sequences is by no means a direct one-to-one relationship. The following brief review of some basic molecular biology will explain this.

The DNA of most living cells contains sequences which code for proteins (including enzymes), transfer RNA (or tRNA, representing a family of molecules involved in protein synthesis) and ribosomal RNA (or rRNA, a major constituent of ribosomes, which are also components of the protein-making apparatus of cells). In the DNA of higher organisms such as eucaryotes, these types of sequences represent only a surprisingly few percent of the total DNA. The rest of the eucaryotic DNA is presumably regulatory sequences, intron fillers between gene fragments, and, above all, great quantities of repetitive DNA. (The latter is often labeled "junk" DNA for its lack of an apparent biological role.) How would the respective biological functions of these DNA classes be reflected in the nucleotide-composition patterns of their DNA sequences? Can we look for some H curve traits that might bear the special trademarks of particular protein classes of the tRNA family or of the rRNA family? Let us start with the two RNA categories whose potentials in this respect are lower than those of protein-coding DNA. The coding sequences for the small tRNA molecules might span only about a few hundred nucleotides, and H curves are not the best means for scrutinizing such short-range DNA sequences. Also, the differences among the dozens of tRNA molecules are rather small, and, furthermore, some of their individual characteristics are not even coded directly in the DNA but are developed after the initial synthesis of tRNA precursors. Thus a systematic study of the H curves of various tRNA coding genes would not be expected to reveal many new insights associated with their "DNA-world" image.

Most of the rRNA molecules are much larger than the tRNAs, and the few varieties that exist are dissimilar to one another. Their biological activity is shared with some specific proteins within the ribosomes. Unfortunately, however, we do not yet know the detailed mechanism of their biochemical activity. Nevertheless, the study of the H curves of DNA regions coding for rRNA precursors might give some interesting results. It is possible that the shapes of their H curves will be found to directly reflect their unique, albeit not well-understood, biological function.

Overwhelming parts of the coding regions of DNA molecules carry blueprints for the synthesis of protein molecules. The repertory of biochemical roles that proteins play in living cells is infinitely more varied than those of tRNA or rRNA molecules. Because an average protein is much larger than most RNA molecules and because proteins are coded on the DNA in nucleotide triplets, the long regions of DNA sequences that code for particular proteins are excellent candidates for H curve analysis. However, there is a major stumbling block in all efforts to look for characteristic H curve features associated with a particular protein (or protein-fragment) function: Because of the redundancy of the genetic code which allows several nucleotide triplets to represent the same amino-acid building block, there are very many ways a particular protein sequence can be encoded on the DNA molecule. Furthermore, there is a second level of redundancy due to the fact that often the biological function of a protein segment can be also accomplished by a related family of other amino-acid sequences. In such a family of functionally equivalent protein sequences there are frequent substitutions selected from sets of operationally identical amino acids. For example, there could be substitutions of one particular hydrophobic amino acid by others of similar kind, a polar amino acid by another polar one, and so on. It is because of this doubly redundant relationship that each protein, or its functional regions (an active site, a structurally important bend of the chain, a domain to be fitted across a membrane, a helix-loop-helix motive, etc.), can be coded for by an astronomical number of related DNA sequences. We have calculated the number of such DNA sequences for the putative biological function associated with the short fragment (amino acids 75 - 91) of the lambda repressor protein which has been extensively studied for all possible amino-acid substitutions [25]. In this protein segment of 17 residues, each amino acid can be substituted on the average by 6 other amino acids without significant alteration of its biochemical activity. According to our calculations (which conservatively assumed that not only single-residue but also multi-residue substitutions would be also possible), there are approximately 10^{20} synonymous DNA sequences, all of which might be translated to play the biological role of this protein fragment [26].

This argument suggests that in the world of DNA sequences a protein fragment associated with a given biological function should be envisioned as a very large set of possible coding DNA sequences and not as a single one. The handling, survey, and analysis of such huge sets of DNA *letter* sequences is, of course, utterly impractical even for very large computers. Fortunately, however, even very large sets of DNA sequences can be displayed together in their H curve form. We designate such sets of related H curves as H curve "bundles."

Figure 4.11 illustrates one of the results of our preliminary work on these types of H curve bundles [27].

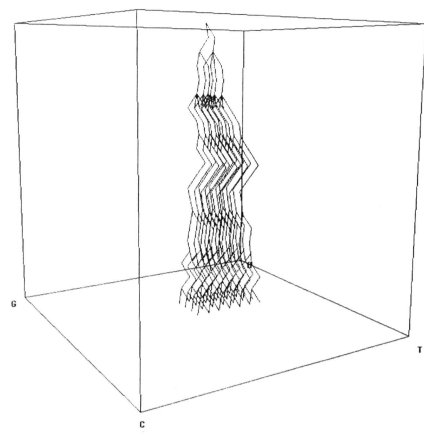

Figure 4.11
Three-dimensional plot of an H curve bundle generated on a Silicon Graphics IRIS/4D graphic workstation representing all the H curves that would code for the heptapeptide Ser-Thr-Tyr-Trp-Asn-Gln-Cys. The 5' ends of all the curves depicted were made to originate from the same point at the top of the reference box.

We are currently experimenting with various visualization approaches for H curve bundles. We would like to optimize the 3D graphic presentation of the multitude of H curve tracks that would best reflect the distinguishing characteristics of these novel expressions of unique biological function. Another current concern is the handling of H curve bunches of much longer polyamino acids such as full protein sequences which consist of a very large number of branching individual H curve tracks. Specialized algorithms and parallelized computations on suitable mini-supercomputers are being applied.

Viewing biological functionality from the peculiar perspective of DNA sequences should appear particularly appropriate from an evolutionary point of view. One might imagine, for instance, that life is controlled by DNA demons inside all living cells and that the prime concern of these demons is the prosperity and propagation of the precious DNA molecules in their charge. The demons work towards their goal by constantly tinkering with the DNA sequences. This results in some sequence alterations. If a certain sequence modification is a successful one, the DNA molecules will survive and prosper; if it is not, they will not. These demons are not skilled organic chemists, however. They are only experts in the art of modification of the metabolic machinery of the cell by a trial-and-error approach. (And they certainly have a decent track record considering the evolutionary history of the last few hundred million years!) The demons have no idea why or how their DNA-sequence concoctions work. We might ask then: How do these demons view DNA sequences? (Certainly not as starting points for the following chain of events: initiation by promoter, transcription, RNA processing, translation, protein modifications, assembly of a catalytic site, acceleration of a biochemical reaction, etc.) They would most likely view DNA sequences in terms of certain large sequence categories which tend to result in successful improvements. Of course, the concept of considering together and simultaneously visualizing such sequence categories is exactly what H curve bundles are all about.

4.8 CONCLUSIONS

In the near future the proliferation of nucleotide-sequence data on massive DNA molecules is expected to create a demand for a suitable graphic metaphor which could concisely represent the very great volume of fundamental genetic information such data embody. An obvious approach would be to use computers with advanced graphics capabilities to create DNA-sequence models which could be manipulated according to the needs of researchers analyzing the data. The method of H curves originated in the author's laboratory appears to offer a suitable starting point to develop such a metaphor. The DNA-conduit paradigm elaborated in this section is an illustration of how such a computer-graphics device could be created for DNA sequences of millions of nucleotides.

There are additional reasons why such a computer-based graphic metaphor of huge DNA sequences will be necessary. The rate of accumulation of DNA sequences (hastened by such efforts as the Human Genome Initiative) will significantly surpass the rate of definitive functional analysis of DNA fragments already sequenced. The desire to explore these proliferating biological *terra incognita* represented by such sequences should create a strong need for suitable methods of global survey, representation, and manipulation.

In the future, most molecular biologists will have access to powerful computer workstations and fast data communication networks. In such a milieu, even massive DNA sequences stored in databases could be quickly accessed and researched by remote users. Under these circumstances, the display, analysis, and various manipulations of a good graphic DNA-sequence metaphor representing several millions of nucleotides would become a realistic possibility.

An additional useful benefit of H curves is related to the fact that they represent DNA-sequence information in an easily superimposable manner. Consequently, it is feasible to represent in a small display area *all* the DNA sequences that could equivalently code for a certain biological function as H curve bundles. Thus, large DNA-sequence sets representing functional/structural motives or even full proteins could be studied and compared in this manner.

ACKNOWLEDGMENTS

The author is indebted to Gabor Varga, Troy Carroll, Matt Delcambre, Joel Welling, Hwa Lim, Hugh Nickolas, Guenther Schoellmann, Richard Hart, Mark Benard, and Jeff Mandel for their invaluable help. Thanks are also due to James LaGuardia for his correction of the manuscript. The work was supported in part by a grant from the Pittsburgh Supercomputing Center (No. DMB900089P, NIH Division of Research Resources cooperative agreement 1 P41 RR06009-01) and from an NSF/LaSER grant (1990-RFAP-13) of the Board of Regents, State of Louisiana. The help and cooperation of the Tulane Computer Services is also gratefully acknowledged.

REFERENCES

[1] S.-I. Aota, T. Gojobori, F. Ishibashi, T. Maruyama, and T. Ikemura, Codon Usage Tabulated from the GenBank Genetic Sequence Data, *Nucleic Acids Res.* **16** (supplement), r315-r323 (1988).

[2] R. Grantham, P. Gautier, M. Gouy, B. Mercier, and A. Pave, Codon Catalog Usage and the Genome Hypothesis, *Nucleic Acids Res.* **8**(1), 49-63 (1980).

[3] T. Ikemura, Codon Usage and tRNA Content in Unicellular and Multicellular Organisms, *Mol. Biol. Evol.* **2**(1), 13-34 (1985).

[4] M.S. Chee, A.T. Bankier, S. Beck, R. Bohni, C.M. Brown, R. Cerni, T. Horsnell, C.A. Hutchison, 3rd, T. Kouzarides, J.A. Martignetti, E. Preddie, S.C. Satchwell, P. Tomlinson, K.M. Weston, and B.G. Barrell, Analysis of the Protein-Coding Content of the Sequence of Human Cytomegalovirus Strain AD169, *Curr. Top. Microbiol. Immunol.* **154**, 125-169 (1990).

[5] G. Myers, A.B. Rabson, J.A. Berzofsky, T.F. Smith, and F. Wong-Staal, *Human Retroviruses and AIDS, 1990*, Los Alamos National Laboratory (1990).

[6] E. Hamori, Low Resolution H Curve of the 48,502 Nucleotides-Long Lambda Genome, *Gene. Anal. Tech.* **1**, 69-74 (1984).

[7] E. Hamori, Novel DNA Sequence Representations, *Nature* **314**, 585-586 (1985).

[8] E. Hamori and J. Ruskin, H Curves, Novel Method of Representation of Nucleotide Series Especially Suited for Long DNA Sequences, *J. Biol. Chem.* **258**(2), 1318-1327 (1983).

[9] E. Hamori, Graphic Representation of Long DNA Sequenes by the Method of H Curves, Current Results and Future Aspects, *BioTechniques* **7**, 710-720 (1989).

[10] E.R. Tufte, *The Visual Display of Quantitative Information*, Graphics Press (1983).

[11] E.R. Tufte, *Envisioning Information*, Graphics Press (1990).

[12] E. Hamori, G. Varga, and J. LaGuardia, HYLAS: Program for Generating H Curves (Abstract 3-Dimensional Representations of Long DNA Sequences), *CABIOS* **5**, 263-269 (1989).

[13] F. Sanger, A.R. Coulson, G.F. Hong, D.F. Hill, and G.B. Petersen, Nucleotide Sequence of Bacteriophage Lambda DNA, *J. Mol. Biol.* **162**, 729-773 (1982).

[14] D.I. Daniels, F. Sanger, and A.R. Coulson, *Cold Spring Harbor Symp. Quant. Biol.* **4712**, 1009-1024 (1982).

[15] E. Hamori and G. Varga, Use of H Curves in Searches for DNA Sequences Which Code for Overlapping Peptide Genes, *FASEB J.* **3**(3), A331 (1989).

[16] E. Hamori, Long Range Nucleotide-Composition Patterns in Fully-Sequenced Genomes, *Biol. Chem. Hoppe-Seyler* **367S**, 226 (1986).

[17] M.G.-S. Lee, C. Loomis, and N.J. Cowan, Sequence of an Expressed Human Beta-Tubulin Gene Containing Ten Alu Family Members, *Nucleic Acids Res.* **12**(14), 5823-5836 (1984).

[18] E. Hamori and G. Varga, DNA sequence (H) Curves of the Human Immunodeficiency Virus 1 and Some Related Viral Genomes, *DNA* **7**(5), 371-378 (1988).

[19] R. Baer, A.T. Bankier, M.D. Biggin, P.L. Deininger, P.J. Farell, T.J. Gibson, G.Hatfull, G.S. Hudson, S.C. Satchwell, C. Seguin, P.S. Tuffnell, and B.G. Barell, DNA Sequence and Expression of the B95-8 Epstein-Barr Virus Genome, *Nature* **310**, 207-211 (1984).

[20] M.P. Cranage, G.L. Smith, S.E. Bell, H. Hart, C. Brown, A.T. Bankier, P. Tomlinson, B.G. Barrell, and, T. C. Minson, Identification and Expression of a Human Cytomegalovirus Glycoprotein with Homology to the Epstein-Barr Virus BXLF2 Product, Variacella-Zoster Virus GpIII, and Herpes Simplex Virus Type 1 Glycoprotein H, *J. of Virology* **62**, 1416-1422 (1988).

[21] A. Razzaque, N. Jahan, D. McWeeney, R.J. Jariwalla, C. Jones, J. Brady, and L.J. Rosenthal, Localization and DNA Sequence Analysis of the Transforming Domain (mtrII) of Human Cytomegalovirus, *Proc. Natl. Acad. Sci. USA* **85**, 5709-5713 (1988).

[22] G. von Heijne, *Sequence Analysis in Molecular Biology*, Academic Press, New York (1987).

[23] P.M. Sharp, What Can AIDS Virus Codon Usage Tell Us?, *Nature* **324**, 114 (1986).

[24] R. Staden, Graphic Methods to Determine the Function of Nucleic Acid Sequences, *Nucleic Acids Res.* **12**(1), 521-539 (1984).

[25] J.U. Bowie, J.F. Reidhaar-Olson, W.A. Lim, and R.T. Sauer, Deciphering the Message in Protein Sequences: Tolerance to Amino Acid Substitutions, *Science* **247**, 1306-1310 (1990).

[26] E. Hamori, G. Varga, M.T. Carroll, and H.A. Lim, Protein Fragment with a Functional Role Viewed as the Sum Total of all DNA Sequences Coding for the Same Function, *FASEB J.* **5**(6), A1536 (1991).

[27] M.T. Carroll, G. Varga, H.A. Lim, and E. Hamori, Novel Graphic Representation of Proteins, in *Proceedings of the Novosibirsk Conference on Modelling and Computer Methods in Molecular Biology and Genetics* USSR Academy of Science Abstracts Supplement, N.A. Kolchanov (ed.), p. 6 (1990).

Visualizing Droplet Coalescence Phenomena

Paul Meakin
Central Research and Development
The Du Pont Company
Wilmington, Delaware

Clifford A. Pickover
IBM Thomas J. Watson Research Center
Yorktown Heights, New York

Fereydoon Family
Department of Physics
Emory University
Atlanta, Georgia

5.1 INTRODUCTION

During the past few years, a renewed interest has developed in processes in which the coalescence of fluid droplets plays a major role. Recently, computer graphics has proved useful for visualizing and understanding models of coalescence and growth. There are several reasons for interest in these subjects, including the following.

1. Droplet coalescence is observed in many familiar and important processes such as food preparation, the condensation of water vapor (and a wide variety of other materials) on relatively cold surfaces, and the deposition of liquid droplets (rain and fog, for example) onto a variety of surfaces.

2. The geometry and evolution of the patterns formed by many processes involving droplet coalescence can be described in terms of simple scaling models.

3. The patterns formed by droplet growth, deposition, and coalescence are often quite beautiful. A familiar example is the decoration of leaves and spider's webs by water droplets.

These three aspects of droplet coalescence have stimulated interest from technological, scientific, and aesthetic points of view respectively. Recent developments concerned primarily with the scientific and technological aspects of droplet coalescence have been surveyed in several recent reviews [1-5]. Here we are concerned primarily with the scientific and aesthetic aspects.

In many systems, fluid droplets have an approximately spherical shape, and no essential information concerning their behavior is lost if we consider them to be spherical. This makes both theoretical work and computer simulations with graphics quite attractive. Of course, the spherical droplet approximation breaks down under a variety of conditions (particularly if the droplets become too large or have been deposited on rough or dirty surfaces with different advancing and receding contact angles).

The beauty of droplet coalescence patterns is not unrelated to their scaling symmetry, which is the focus of current scientific interest. The main objective of this chapter is to show how computer simulations and computer graphics can be used to develop a better understanding of important natural phenomena. By means of high-quality computer graphics, we will attempt to show how computer simulations can be used to mimic (and in some respects enhance) the complex patterns generated by natural phenomena.

In all of the models illustrated here, it is assumed that the droplets have a hyperspherical shape or a hyperspherical "cap" shape with a constant contact angle. For such droplets, the droplet radius r is related to its mass or size s as follows:

$$r \sim s^{1/D} \tag{5.1}$$

where D is the dimensionality of the droplet. It is assumed that the droplets are embedded in a d-dimensional space or sit on a d-dimensional substrate. In general, D and d can take on essentially any value, but we will show only results for $d = 1$, 2, or 3 and integer values of D. If two (or more) droplets contact each other, they are immediately coalesced into a single drop with conservation of both the total mass and center of mass of the coalescing group. This means that the radius r of the coalesced droplet is given by

$$r^D = \sum_{i=1}^{n} r_i^D \tag{5.2}$$

where r_i is the radius of the ith droplet and n is the number of component droplets. Similarly, the position $\mathbf{x}$ of the center of the coalesced droplet is given by

$$\mathbf{x} = \sum_{i=1}^{n} \mathbf{x}_i s_i / S \tag{5.3}$$

where $\mathbf{x}_i$ is the (vector) position of the center of the ith component droplet and s_i is its size. Here, S is the size of the combined droplet.

5.2 DROPLET GROWTH AND COALESCENCE MODELS

Perhaps the most simple droplet coalescence models are those in which it is assumed that droplets grow and coalesce on contact without renucleation in the spaces between the growing droplets. In one example of this type of model [1-7], the individual droplets are assumed to grow according to the power law

$$\frac{dr}{dt} \sim r^\omega \quad \text{or} \quad \frac{dr}{dt} = k r^\omega \tag{5.4}$$

where r is the droplet radius. At the start of a simulation, N_0 droplets with a diameter of d_0 (or radius r_0) are deposited (placed) randomly in a system of size L^d, avoiding overlap between any of the droplets. In these droplet growth and coalescence models, individual droplets grow according to (5.4) until they contact other droplets and coalesce. In practice, the radii of all of the droplets are increased in small steps so that

$$r' = \left(r^\xi + \delta\xi\right)^{1/\xi} \tag{5.5}$$

where r' is the new droplet radius, ξ is $1 - \omega$, and δ is a small number. The value of δ is adjusted so that an increase in δ by a factor of at least 2 leads to changes in the simulation results that are smaller than the statistical uncertainties. The algorithm consists of a sequence of dilation (5.5) and coalescence (if overlaps are found) steps. Periodic boundary conditions were used in all of the simulations illustrated in this chapter.

Figure 5.1 shows some droplet patterns obtained from simulations carried out with three-dimensional droplets on a two-dimensional substrate (it is useful to think of hemispherical caps growing on a planar substrate). At the start of the simulation, 20,000 droplets with a diameter $d_0 = 1.5$ were placed randomly (avoiding overlap) on a substrate with an area of 512×512. A value of 0 was used for the growth exponent, ω, in (5.4). In this figure, the droplet patterns are shown on length scales that are proportional to $S^{1/D}(R_s)$. This figure demonstrates the scaling symmetry associated with the droplet patterns and indicates directly that the droplet size distribution can be represented by the scaling form

$$N_s(t) \sim s^{-\theta} f(s/S(t)) \tag{5.6}$$

where $N_s(t)$ is the number of droplets of size s at time t and $S(t)$ is the mean droplet size defined as

$$S(t) = \frac{\int_1^\infty s^2 N_s(t) ds}{\int_1^\infty s N_s(t) ds} \tag{5.7}$$

where the droplet sizes are measured in units of the size (mass) of a droplet with diameter d_0. The scaling form [8] expressed in (5.6) describes very well the size distributions in a wide range of systems undergoing aggregation, coalescence, fragmentation, and so on. In the case of the droplet growth and coalescence model described above, it can be shown, using simple scaling arguments [5,7], that the exponent θ in (5.6) is given by

$$\theta = \frac{D + d}{D} \tag{5.8}$$

Similar simulations have been carried out using various combinations of the parameters D, d and ω. For $d = 1$ or 2, the simple "black and white" laser printer graphics illustrated in Figure 5.1 is quite adequate. However, for $d = 3$ it is more difficult to obtain an adequate visualization of the spatial droplet distribution. Figure 5.2 shows a droplet pattern generated in a simulation of the growth and

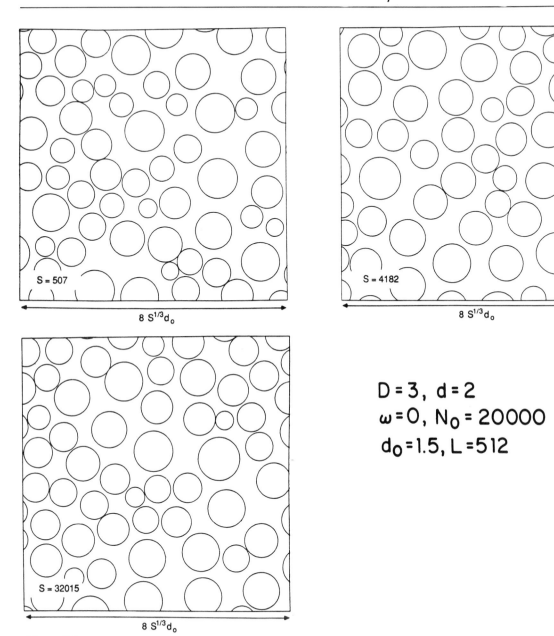

$$D = 3, \ d = 2$$
$$\omega = 0, \ N_0 = 20000$$
$$d_0 = 1.5, \ L = 512$$

Figure 5.1
Three stages in a simulation of droplet growth and coalescence with $D = 3$, $D = 2$, and $\omega = 0$. Each part of the figure is shown on a scale proportional to $S^{1/D}(D_s)$ to illustrate the scaling symmetry.

(a)

Figure 5.2
A droplet pattern generated
using a model for the growth
and coalescence of three
dimensional droplets in a three
dimensional substrate. (a)
Image generated using a
specular reflection model in
which each sphere is treated
independently. (b) Image
obtained using a ray tracing
algorithm for the complete
system to produce shadows and
reflections. See insert for color
representation.

(b)

coalescence of three-dimensional droplets in a three-dimensional space
($D = 3$ and $d = 3$) with a growth exponent $\omega = 0$. At the start of the
simulation, 10^5 droplets with a radius of 0.9 were placed randomly
(avoiding overlap) in a cubic box of size $128 \times 128 \times 128$. Figure 5.2
shows the system at the stage at which the number of droplets (N) has
decreased to 499 and the mean droplet size (S) has reached a value
of 15,381 (in units of the mass of the initial droplets with a radius of
0.9).

Here, we present both ray-traced and non-ray-traced models for
both visual clarity and aesthetics. Figure 5.2a was produced by general-
purpose, custom software running on an IBM RISC System/6000 allowing us to interactively light and rotate the three-dimensional collection of spheres. For Figure 5.2a, lighting of the shapes was applied
on a primitive-by-primitive basis; no interactions between objects such
as shadows or reflections were defined. The reflectance calculation is
applied at points on the spheres being lit and shaded, and produces
color at these points. Input to the reflectance calculation includes (a)
the position on the primitive at which the reflectance equation is being
applied, (b) the reflectance normal, (c) diffuse color at that position,
(d) the set of light source representations, and (e) the eye point. Three
lights were used (red, white, and green), and the lights were rotated in
realtime, using a mouse, to produce the desired final graphical effects.
Specular reflections produced the highlights on the shiny surfaces. As
is traditional, the intensity of specular reflections, unlike diffuse reflections, is highly dependent on the viewing angle of the observer.
The specular exponent used to control the shininess was about 100.

Powerful graphics workstations can be used to rotate the collection of spheres with only a few seconds pause between images. The
required computations for an animated sequence of droplets could
take hours on traditional high-powered mainframe computers. Figure 5.2b was produced by standard ray-tracing techniques in order to
produce shadows and reflections. It required around an hour of CPU
time to compute on a RISC System/6000.

For a single droplet growing according to (5.4), the asymptotic
time dependence of the droplet radius $r(t)$ and size $s(t)$ are given by

$$r(t) \sim t^{1/(1-\omega)} \tag{5.9}$$

and

$$s(t) \sim t^{D/(1-\omega)} \tag{5.10}$$

The same algebraic forms describe the growth of the mean droplet
size $S(t)$ [equation (5.7)] and the mean droplet radius $R_s(t)$ in the

corresponding droplet growth and coalescence models [5,7]

$$R_s(t) \sim t^{1/(1-\omega)} \tag{5.11}$$

and

$$S(t) \sim t^{D/(1-\omega)} \tag{5.12}$$

However, the "prefactors" or "amplitudes" are renormalized via the coalescence process. This is not true for all droplet growth and coalescence models [2]. A more detailed discussion of the various droplet growth and coalescence models [5] would be beyond the scope of this chapter. A clear discussion of the principal growth and coalescence models can be found in the thesis of Galvin [4].

5.3 DROPLET DEPOSITION AND COALESCENCE

In the droplet deposition and coalescence model [1,5,9], D-dimensional (hyperspherical) droplets are "dropped" randomly, one at a time, onto a d-dimensional substrate. This model can be considered to represent phenomena in which macroscopic droplets are deposited onto a smooth substrate, but it also represents quite well the process of droplet growth and coalescence with renucleation of droplets on those regions that are left uncovered as a result of the coalescence of large droplets [2,10]. Figure 5.3 shows an example of such a process in which tin vapor was deposited onto a sapphire surface [9]. Figure 5.4 was obtained using a model in which three-dimensional droplets with a diameter $d_0 = 1.5$ were dropped randomly onto a substrate of area 512×512 (with periodic boundary conditions). The droplets were assumed to have a diameter of d_0 after deposition onto the surface, and all the droplets were assumed to have the same "spherical cap" shape. The pattern shown in the bottom left corner of Figure 5.4 resembles quite closely the experimental pattern shown in Figure 5.3. In the latter stages of a simulation of this type, redeposition occurs on the areas that have been cleared by the coalescence of very large droplets. The entire deposition and coalescence process is repeated in miniature on the cleared areas, and it is apparent that for a sufficiently large system an indefinite hierarchy could, in principle, develop.

Figure 5.4 shows four stages in a simulation of the deposition of three dimensional droplets onto a two dimensional substrate. The four parts of this figure are shown on scales proportional to $S^{1/3}$ (i.e., $S^{1/D}$

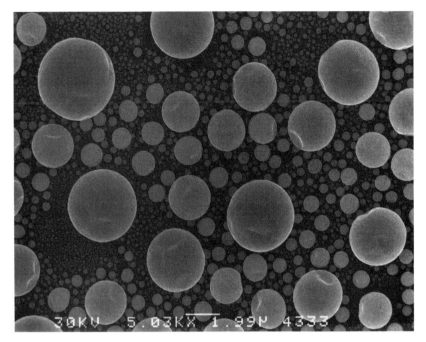

Figure 5.3
A pattern formed by the low-pressure deposition of tin vapor onto a sapphire surface held at a temperature near to the melting point of tin (232°C) [9]. See insert for color representation.

or R_s). If the small droplets are ignored, the patterns of large droplets look similar at all four stages (particularly in the later stages which are well into the scaling regime). This suggests that the distribution of large droplets can be described in terms of the scaling form given in (5.6). Again, simple scaling arguments indicate that the exponent θ in (5.6) is given by (5.8) ($\theta = (D + d)/D$), and this has been confirmed by extensive computer simulations. In fact, the entire droplet size distribution can be scaled in this manner [9].

Although Figure 5.4 provides a quite good representation of the evolution of the droplet pattern for the case $D = 3$ and $d = 2$, a much more realistic picture can be obtained using more advanced color graphics. Figure 5.5 shows the droplet pattern obtained from a simulation in which 2×10^7 droplets each with a diameter $d_0 = 1.5$ were deposited onto an area of 640×640.

Again, two representations of this simulation are shown. Figure 5.5a shows a figure generated with specular reflection at the individual droplets, and Figure 5.5b shows a figure generated using a full ray tracing algorithm which includes reflection, refraction, and shadowing effects.

Simulations have been carried out for a variety of combinations

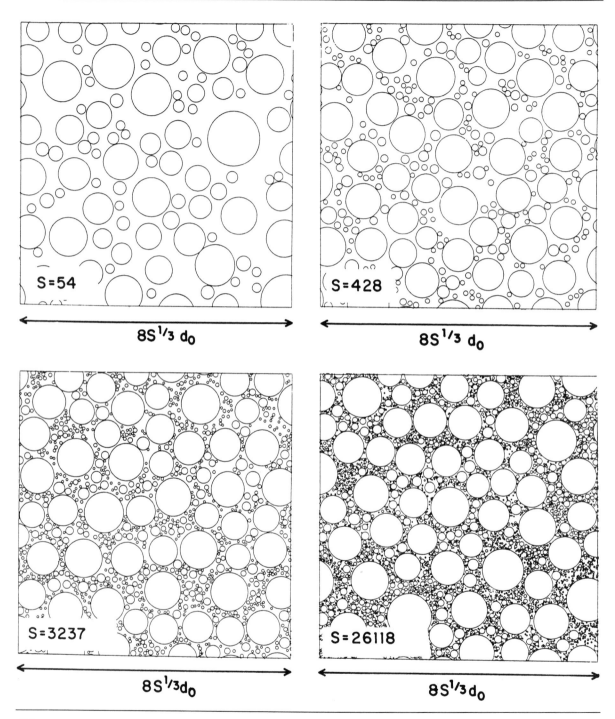

(a)

(b)

Figure 5.5

A droplet configuration generated using a model for the random deposition and prompt coalescence of three-dimensional droplets on a flat two-dimensional substrate. (a) Image obtained using a specular reflection algorithm without optical interaction between droplets. (b) Image obtained using a full ray tracing model that includes optical interactions between droplets. See insert for color representation.

of D and d. The case $D = 3$ and $d = 1$ does not correspond to the deposition of droplets onto a fiber (for example, the deposition of fog droplets onto a spider's web) since the model implicitly assumes the deposition of droplets from a $(d + 1)$-dimensional space onto a d-dimensional substrate. However, Figure 5.6 shows the results of a simulation of deposition onto a line from a three-dimensional space. In this case, the large droplets on the line capture a larger fraction of the small deposited droplets than they do in the $D = 3$, $d = 1$ version of the "standard" deposition and coalescence model. As in the case of the growth and coalescence model, recognition of the appropriate scaling symmetry leads to a quantitative description of the process.

For the deposition of D-dimensional droplets onto a d-dimensional substrate from a $(d + 1)$-dimensional space, the mean droplet size $S(t)$ grows algebraically,

$$S(t) \sim t^z \tag{5.13}$$

Figure 5.4

(Opposite page.) Four stages in a simulation of droplet deposition and coalescence ($D = 3$, $d = 2$). The patterns are shown on length scales that are proportional to $s^{1/D}$ to illustrate the scaling properties associated with this class of models.

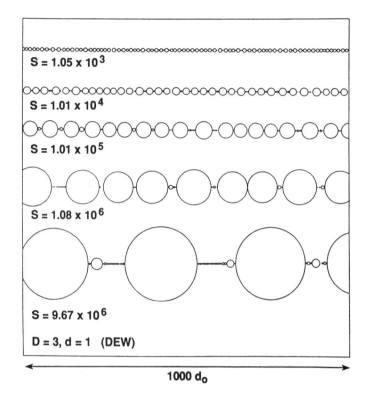

Figure 5.6
Deposition of three-dimensional droplets onto one-dimensional substrates. This figure obtained by simulated deposition from a three-dimensional space corresponding to the growth of droplets on a fiber (deposition of dew on a spider's web, for example).

and the exponent z is given by reference 9:

$$z = \frac{D}{(D - d)} \qquad (5.14)$$

This result can be derived using simple scaling arguments and is confirmed by computer simulations [9]. Similarly, the asymptotic decrease in the number of large droplets is given by

$$N_L(t) \sim t^{-z''} \qquad (5.15)$$

where the exponent z'' is given by

$$z'' = \frac{d}{(D - d)} \qquad (5.16)$$

The total number of droplets $N(t)$ also decay algebraically,

$$N(t) \sim t^{-z'} \qquad (5.17)$$

and the exponent z' is related to the exponent z [equations (5.13) and (5.14)] and θ [equations (5.6) and (5.8)] by

$$z' = z(\theta - \tau) \tag{5.18}$$

where τ is the exponent that describes the power law distribution of small droplet sizes:

$$N_s(t) \sim s^{-\tau} g\left(\frac{s}{S(t)}\right) \tag{5.19}$$

The scaling function $g(x)$ in (5.19) has a constant value for $x \ll 1$. So far we do not have a scaling theory for the exponents τ and z'.

5.4 DROPLET DEPOSITION, SLIDING, AND COALESCENCE

The motion of large droplets on an inclined surface is an important process in the breath figure [2,10] or dropwise condensation process that has important implications in heat transfer engineering [11-13].

The most simple way of including droplet sliding is to assume that when a droplet reaches a critical size s_c it slides "down" the substrate, colliding and coalescing with all the droplets that it contacts [13,14]. Because the sliding droplet grows as a result of this process, the path that it sweeps "clean" becomes wider as the droplet moves. The shape of this path depends on the dimensionality D of the droplet. It is assumed, in most cases, that the droplet sliding process is complete and the sliding droplet exits the system before further deposition occurs.

Figure 5.7 shows several $d = 2$ systems just after the first droplet has exceeded the maximum size ($s_{max} = s_c - 1$). These figures were obtained using the particle deposition and coalescence model described above with instantaneous sliding and coalescence of the droplets that exceed a size s_{max} [14]. In a sufficiently large system, the size of the sliding droplets will greatly exceed the characteristic sizes R_s and R_d (the distance between droplets) associated with the distribution of droplets on the substrate. In this limit, the sliding droplet can be considered to be moving and collecting the mass from a system with a uniform surface density $\rho(s_{max})$, where $\rho(s_{max}) \sim (s_{max})^{1/D}$ in the limit $s_{max} \gg s_0$. Based on this idea, it can be shown [5,14] that the distribution of first "avalanche" masses (N_s^1) is given by

$$N_s^1 = s^{-\tau} f(s/S^*) \tag{5.20}$$

where the function $f(x)$ has the form $f(x) =$ const. for $x \ll 1$ and

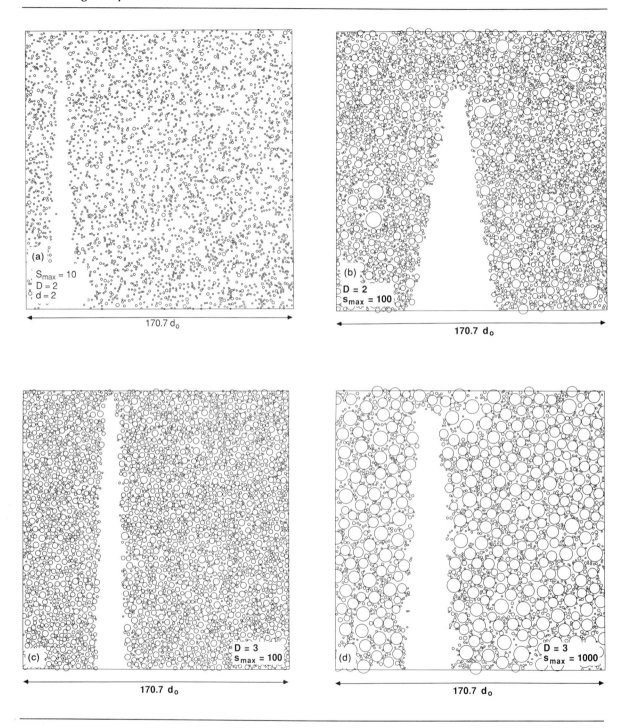

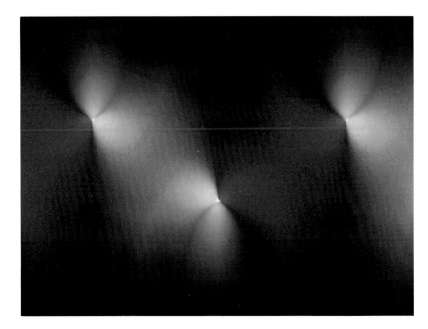

Figure 1.3
Page 13

Figure 1.6
Page 18

For complete figure caption, see the page number(s) indicated.

Figure 1.9a

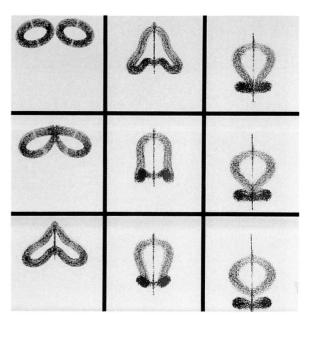

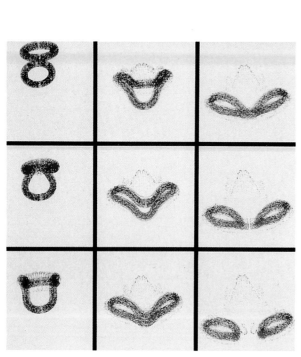

Figure 1.9b

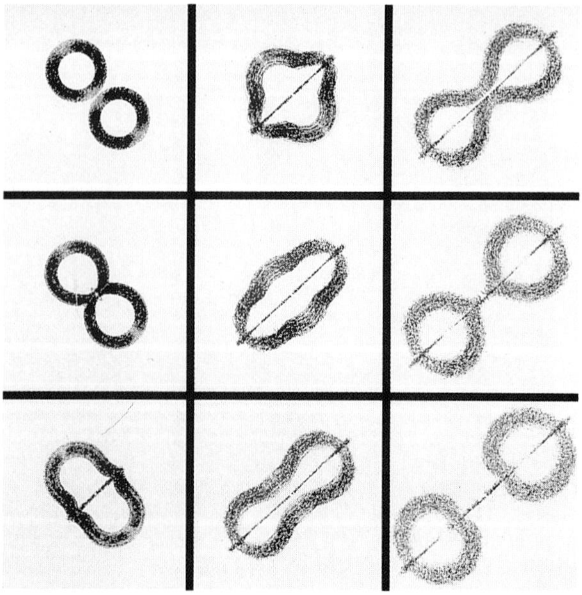

Figure 1.9c
Pages 24–25

For complete figure caption, see the page number(s) indicated.

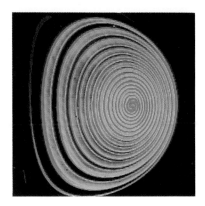

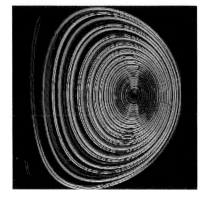

Figure 1.12
Page 27

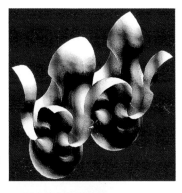

Figure 1.10
Page 25

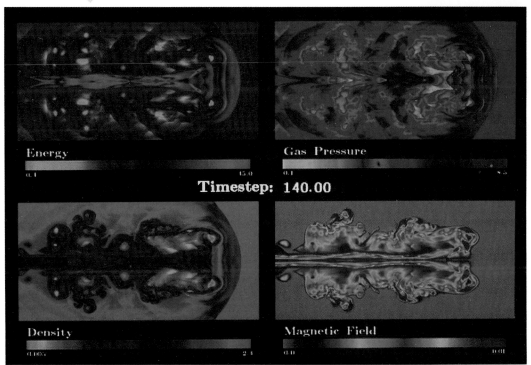

Figure 1.14
Page 29

Figure 1.15
Page 30

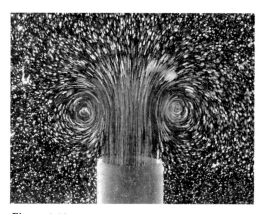

Figure 1.18
Page 33

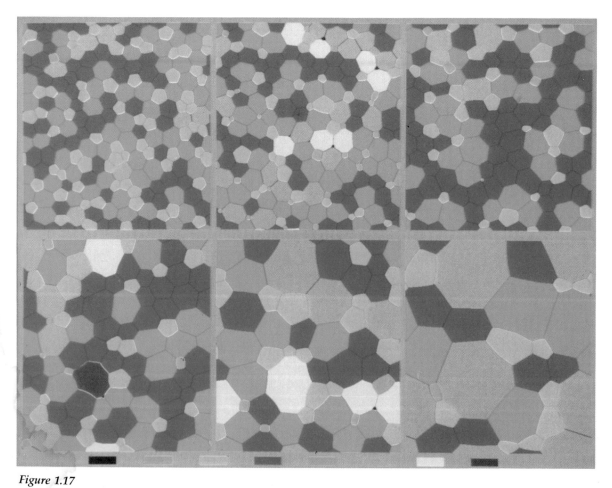

Figure 1.17
Page 33

For complete figure caption, see the page number(s) indicated.

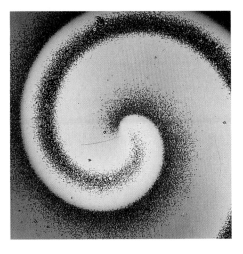

(Left) **Figure 3.1**
Page 71

(Right) **Figure 3.2**
Page 72

(Left) **Figure 3.3**
Page 74

(Right) **Figure 3.5**
Page 76

(Bottom Left)
Figure 3.10
Page 82

(Bottom Right)
Figure 3.13
Page 84

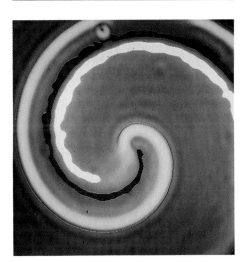

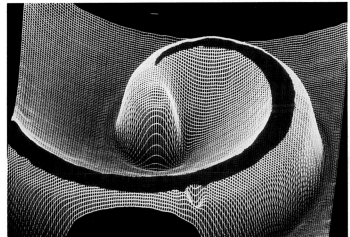

Figure 5.2 a and b
Page 128

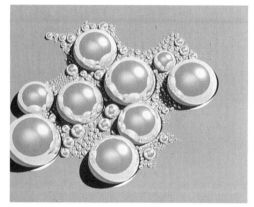

Figure 5.5 a and b
Page 133

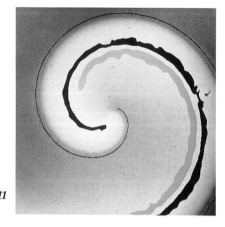
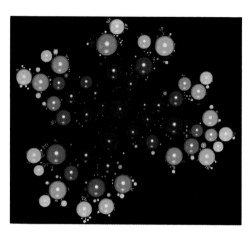

Figure 3.11
Page 83

Figure 5.10
Page 140

For complete figure caption, see the page number(s) indicated.

(a)

(b)

Figure 6.10 a and b Page 170

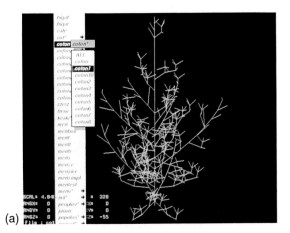

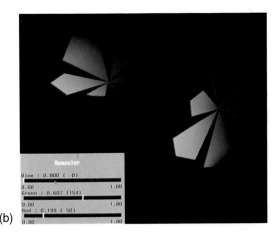

(a)

(b)

(c)

(d)

Figure 6.12 a–d Page 172

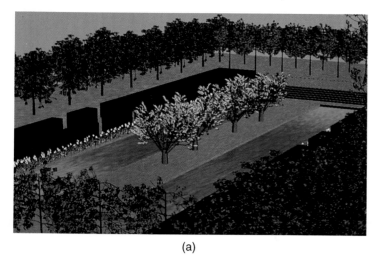

(a)

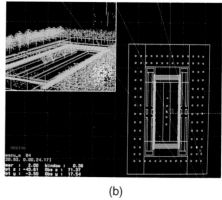

(b)

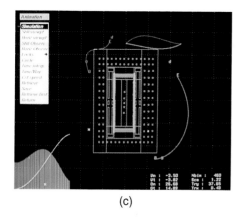

(c)

(d)

Figure 6.13 a–e
Page 173

(e)

For complete figure caption, see the page number(s) indicated.

Figure 6.15 a and b (a)
Page 174

(b)

Figure 6.14 a and b
Page 175

(a)

(b)

Figure 7.2
Page 187

Figure 7.3
Page 187

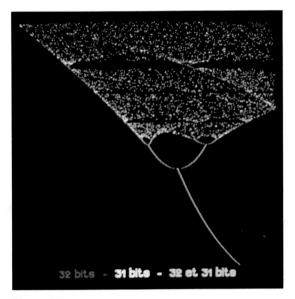

Figure 7.5
Page 188

Figure 7.8
Page 190

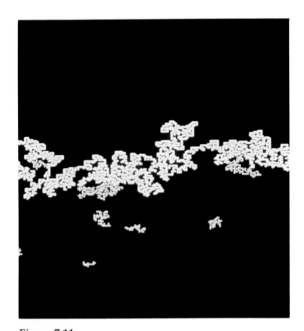

Figure 7.11
Page 193

Figure 7.12
Page 193

For complete figure caption, see the page number(s) indicated.

Figure 7.14
Page 198

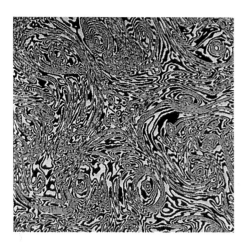
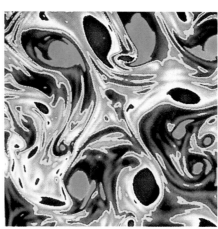

Figure 7.15
Page 203

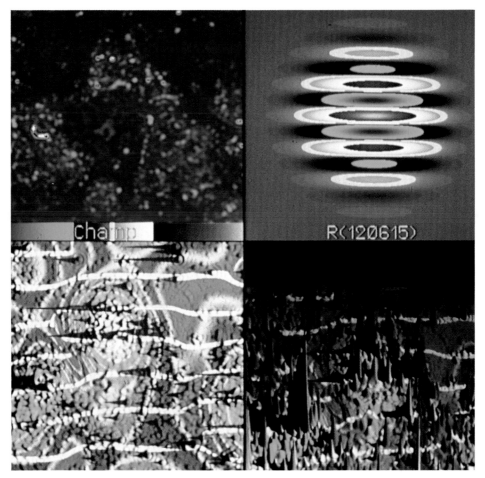

Figure 7.16
Page 203

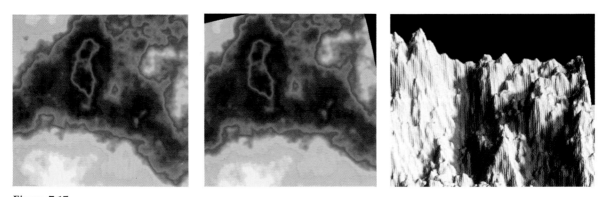

Figure 7.17
Page 204

For complete figure caption, see the page number(s) indicated.

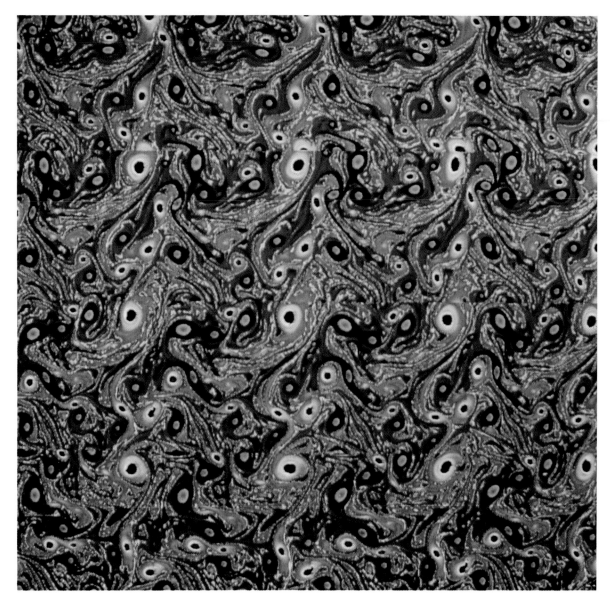

Figure 7.18
Page 205

Figure 7.19
Page 206

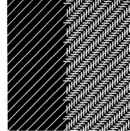

Figure 7.20
Page 206

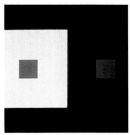

Figure 7.21
Page 207

Figure 7.22
Page 207

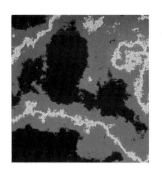

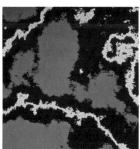

Figure 7.23
Page 208

Figure 7.24
Page 208

(a)   (b)   (c)

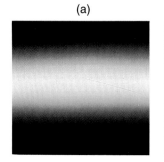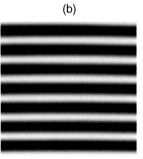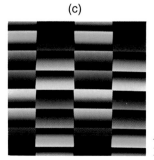

Figure 7.25
Page 209

For complete figure caption, see the page number(s) indicated

Figure 8.7
Page 231

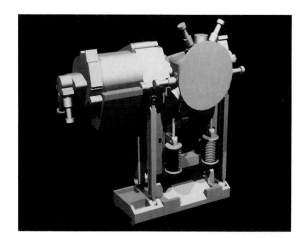

Figure 8.8
Page 232

Figure 8.9
Page 232

Figure 8.11
Page 233

Figure 8.10
Page 233

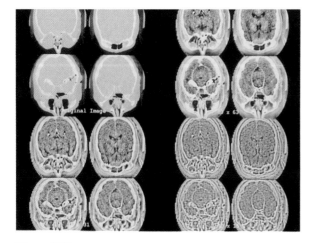

Figure 8.15
Page 236

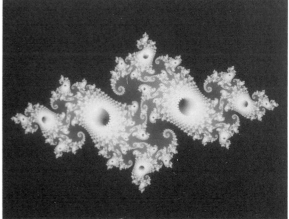

Figure 8.17
Page 237

For complete figure caption, see the page number(s) indicated.

where $f(x)$ decays faster than any power of x for $x \gg 1$. The cut-off S^* is given by

$$S^* \sim (\rho L)^{D/(D-d+1)} \qquad (5.21)$$

and the exponent τ is given by

$$\tau = (d-1)/D \qquad (5.22)$$

In (5.21) ρ represents $\rho(s_{max})$.

If the deposition process is continued after the first avalanche, a steady-state is eventually reached. Both the mean droplet size, $S(t)$, and the number of droplets, $N(t)$, exhibit oscillations before the steady state is reached. Figure 5.8 shows several steady-state configurations obtained from off-lattice simulations with two-dimensional substrates.

An analytic calculation for the steady-state regime has not yet been carried out for $d > 1$. However, computer simulation results indicate that the dependence of the mean droplet size $S(y, \infty)$ and the steady-state mean density $\rho(y, \infty)$ on the distance y from the "top" end of the substrate are both algebraic

$$S(y, \infty) \sim y^{-\omega}$$

$$\rho(y, \infty) \sim y^{-\gamma}$$

while the number density of the droplets has, at most, a logarithmic dependence on y.

The simple scenario described above is not expected to provide a realistic description of most real systems. For example, some of the complications associated with the deposition of water droplets on inclined glass sheets are described by Janosi and Horvath [15].

Figure 5.9 shows results from a larger-scale simulation carried out for the $D = 3$, $d = 2$ case. The simulation was carried out by depositing

Figure 5.7
(Opposite page.) Results obtained from the two-dimensional off-lattice model for droplet deposition, coalecence, and sliding. Here, several systems immediately after the first "avalanche" has occurred are shown. D is the dimensionality of the droplets, and s_{max} is the maximum droplet size ($s_c = s_{max} + 1$).

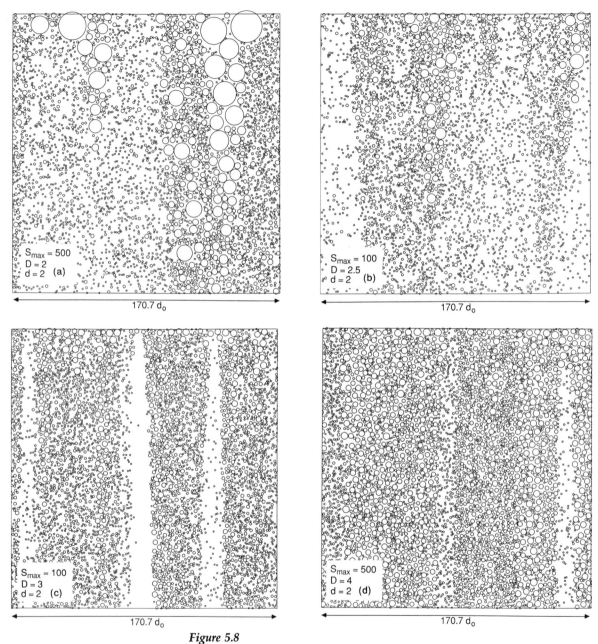

Figure 5.8
Typical steady-state droplet configurations obtained using different values for
the droplet dimensionality (D) and maximum droplet size (s_{max}) with a two-
dimensional substrate.

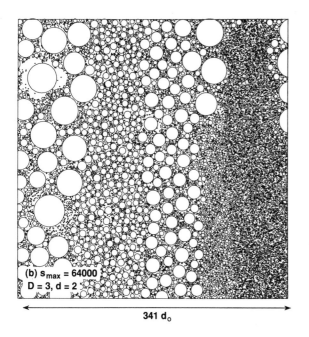

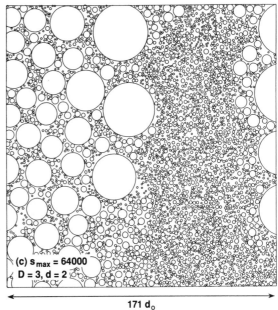

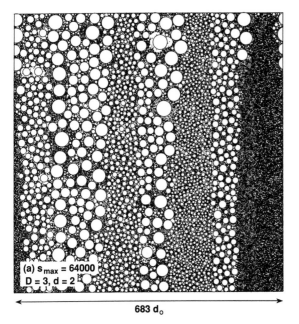

Figure 5.9
A droplet pattern obtained from the droplet deposition, coalescence, and sliding model. (a) The entire pattern obtained from a large-scale simulation for three-dimensional droplets on a two-dimensional substrate. (b) The upper right quadrant of (a). (c) The upper right quadrant of (b).

particles with a diameter $d_0 = 1.5$ on a substrate of size 1024×1024. The sliding threshold mass s_{max} was set to a value of 64,000 s_0 (s_0 is the size or mass of a droplet of diameter d_0). The simulation was continued until a time of $4t^*$, where t^* is the time (number of deposited droplets) at which the first avalanche appears. At this stage the system is near to, but has not quite reached, the steady state. The first avalanche occurs after approximately 10^7 droplets have been deposited. Figure 5.9a shows the entire system at time $t = 4t^*$. Figure 5.9b shows the upper right quadrant of Figure 5.9a, and Figure 5.9c shows the upper right quadrant of Figure 5.9b.

5.5 COALESCENCE PHENOMENA ON COMPLEX SUBSTRATES

While most attention has been focused on experimental studies and computer simulations of coalescence phenomena in and on simple Euclidian substrates, droplet growth and coalescence phenomena often take place in environments that have a complex, disorderly geometry. To illustrate this type of process, we have carried out simulations of droplet growth and coalescence using clusters generated using a two-dimensional diffusion-limited (DLA) model [16] as a substrate. These clusters have a randomly branched structure that can be characterized by a fractal dimensionality [17] of about 1.715.

To generate the droplet pattern shown in Figure 5.10, a 50,000-

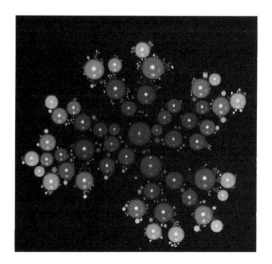

Figure 5.10

A simulation of the growth of three-dimensional droplets on a two-dimensional (off-lattice) DLA cluster. The color scale indicates the stage at which the particles in the underlying DLA cluster were added to the growing cluster. See insert for color representation.

Figure 5.11
Simulated dew drops on a
spider web.

particle two-dimensional off-lattice DLA cluster was first generated
using a simple but efficient algorithm [18]. The centers of each of the
particles (disks with a diameter of 1.6) were used as nucleation sites in
a simulation of droplet growth and coalescence with a growth exponent (ω) of zero. At the start of each simulation, a droplet "embryo"
with a diameter of 0.015 was placed at each of the nucleation sites.
When nucleation sites became exposed as a result of droplet coalescence, droplet "embryos" were placed at the exposed nucleation sites
and growth began anew. Figure 5.10 shows the droplet pattern generated using this model. In this figure, the droplets have been assigned
a color that indicates the mean "age" of the particles in the underlying
DLA cluster that are associated with the individual droplets. Figure
5.11 shows a simulation of dew drops formed on a spider web.

5.6   SUMMARY

This chapter illustrates how computer graphics can be used to present
the results of computer simulations in a clear and concise manner. A
verbal description of Figures 5.4 and 5.7-5.10 would be completely

inadequate. These figures (and others like them) contain many interesting and important details similar to those observed in real systems (Figure 5.3, for example). It is, of course, important to obtain quantitative information from both experiments and simulations, and we have devoted considerable effort to this [1,3,5,6,9,14]. However, a visual comparison between experimental and simulated patterns often provide a fast and convincing assessment of the quality of a computer simulation.

High-quality computer graphics also plays an important role in motivating both the researcher and those who must understand the results of a simulation (colleagues, students, and the public at large). For this reason alone, computer graphics is playing an increasingly important role in scientific research.

Because nature is often beautiful, it is not surprising that images based on simulations are frequently beautiful. Using computer simulations, we can explore systems and conditions that are physically inaccessible. Beautiful and fascinating images are also frequently associated with these simulations.

REFERENCES

[1] F. Family, in *Random Fluctuations and Pattern Growth: Experiments and Models*, H. E. Stanley and N. Ostrowsky (eds.), Kluwer, Dordrecht, p. 345 (1988).

[2] D.H. Fritter, *Breath Figures: The Evolution of Droplet Patterns on Surfaces Through Growth and Coalescence*, Ph.D. Thesis, University of Californi at Los Angeles (1989).

[3] P. Meakin, in *Dynamics and Patterns in Complex Fluids*, A. Onuki and K. Kawasaki (eds.), Springer Verlag, Berlin (1990).

[4] K.P. Galvin, *Growth and Coalescence in Condensation*, Ph.D. Thesis, University of London (1990).

[5] P. Meakin, *Rep. Prog. Phys.* **55**, 157 (1992).

[6] J.W. Rose and L.R. Glicksman, *Int. J. Heat Mass Transfer* **16** 411 (1973).

[7] P. Meakin and F. Family, *J. Phys.* **A22**, L225 (1989).

[8] T. Vicsek and F. Family, *Phys. Rev. Lett.* **52**, 1669 (1984).

[9] F. Family and P. Meakin, *Phys. Rev. Lett.* **61**, 428 (1988). Also: *Phys. Rev.* **A40**, 3836 (1989).

[10] D. Beysens and C.M. Knobler, *Phys. Rev. Lett.* **57**, 1433 (1986).

[11] E. Citakoglu and J.W. Rose, *Int. J. Heat Mass Transfer* **12**, 645 (1986).

[12] R.E. Tower and J.W. Westwater, *Chem. Eng. Prog.* **66**(102), 21 (1970).

[13] N. Fatica and D.N. Katz, *Chem. Eng. Prog.* **45**(11), 661 (1949)

[14] Z. Cheng, S. Redner, P. Meakin, and F. Family, *Phys. Rev.* **A40**, 5922 (1989).

[15] I.M. Janosi and V.K. Horvath, *Phys. Rev.* **A40**, 5232 (1989).

[16] T.A. Witten and L.M. Sander, *Phys. Rev. Lett.* **47**, 1400 (1981).

[17] B.B. Mandelbrot, *The Fractal Geometry of Nature*, W.H. Freeman and Company, New York (1982).

[18] P. Meakin, *J. Phys.* **A18**, L661 (1985).

Computer Simulation of Plant Growth

Philippe de Reffye
CIRAD/Laboratoire de Modélisation
Montpellier Cedex, France

6.1 INTRODUCTION

Nature is a kaleidoscope of shapes, and as we see throughout this book, mathematical formulas can sometimes be used to simulate natural forms. This chapter focuses on the simulation of botanical shapes and also provides a sampling of concepts used in the simulation of plant growth. Many botanical terms are introduced so readers will get a flavor of the kinds of concerns researchers deal with when accurately simulating plant growth using a computer.

The idea that nature and mathematics are inextricably linked is not new – and neither is the application of that idea in computer graphics. A number of published papers have addressed the generation of mathematically derived morphological models for plants. These and other successes provide continuing incentive for more research on the mathematical basis of plant structure.

Few biological fields of investigation have developed the computer-aided design tools often used in the physical sciences. This seems to be due to the difficulty of identifying the mathematical and other parameters of a biological system. Therefore, until recently, botanists and agronomists still lacked necessary tools. A plant's complex architecture remained inaccessible because there was no method of rational analysis with which to measure it; trees were described only part by part (leaves, flowers, bark, etc.). General, imprecise terms such as "weeping habit" and "spherical habit" were used to describe their shape. Plants were only drawn in their totality by artists whose talent had to make up for their lack of scientific observation.

Much progress has been made in structural biology over the past 20 years. In 1970, a school which specialized in morphological descriptions of tree architecture was founded by Prof. F. Halle [1] in Montpellier (France), a city known for its botanical studies for over ten centuries. Halle, along with R.A.A. Oldeman [2], defined the basic concepts of the architectural analysis of trees – that is, the architectural model and "reiteration," which we will discuss later. Botanists of this school developed a rational scientific method for visual representation of trees and were capable of expressing the main elements and growth strategies of trees. Most of F. Halle's team's contributions were essentially of a qualitative nature and were necessary for the elaboration of a consistent mathematical model; nevertheless, these qualitative data were still insufficient for the elaboration of a quantitative modeling of plant growth. However, the goal of such a modeling is very important. Once attained, selection criteria could be defined, pruning and density techniques optimized, the development of trees in a given environment predicted. Industrial applications in landscape architecture (landscapes, parks, gardens) could also make use of three-dimensional computer-generated images.

As alluded to at the beginning of this chapter, computer graphics of trees generated by algorithms already exist. These include model trees generated by specific branching processes [3], graftals [4], paracladial systems [5], fractals [6], combinatorial trees [7], and thin transparent ellipsoids [8].

Although these models are interesting from an algorithmic point of view as well as for their graphic results, they do not have a thorough botanical basis and an experimental approach. Except for the case of small plants which are obviously easier to model [9], the simulated trees created in the past can only be viewed from far away because, when seen from close up, they reveal too many artifacts of the mathematical algorithm which produced them.

In the CIRAD (Centre de Coopération Internationale en Recherche Agronomique pour le Développement) Modeling Laboratory, we have formed a team of 20 researchers specialized in four different fields (botany, agronomy, mathematics, computers), allowing for a quantitative approach to this problem. Our belief is that a well-balanced interaction of these disciplines is necessary to establish the fundamental principles of growth simulation and plant architecture. In this chapter, we present the current state of our investigations.

6.2 BOTANICAL BASIS OF PLANT ARCHITECTURE

Our tree simulation is based on some qualitative data, well summarized by C. Edelin [10], including knowledge acquired by F. Halle and his co-researchers concerning tropical trees.

As background, in the earliest stages of embryonic growth, cell division takes place throughout the body of the infant plant. As the embryo grows older, however, new cell addition becomes restricted to certain parts of the plant body – for example, the "apical meristems" of the root and the shoot. During the remainder of the plant's life, all the primary growth (which involves the elongation of the plant body) will originate in these meristems. In Figure 6.1b, the leafy axis is the result of the functioning of an apical meristem (Figure 6.1a) which produces node-bearing leaves with their own axillary meristems. The stem portion which separates two successive nodes is called an *internode*. It represents the natural unit for measuring branch length.

Meristems exhibit different behaviors and patterns which lead to the specific topological and geometrical characteristics of the leafy axis. For example, Figure 6.1c shows a continuous pattern, and Figure 6.1d shows a rhythmic one. The stem portion which separates two growth cessation points of the stem is referred to as a *growth unit* (G.U.). Branching of the leaf axis can be diffuse, continuous, or rhythmic.

The sexuality can be lateral in the indeterminate growth axes, producing monopodial trees (Figure 6.1e). (With monopodial branching, a branch divides in two at the growth point, and one follows the direction of the main axis while the other goes in a different direction to form a lateral branch.)

If it is terminal in the determinate growth axes, the tree grows by successive stacking of relay axes and becomes a sympodial structure (Figure 6.1f). Some axes tend to grow vertically; they are called *orthotropic*. Other axes tend to grow horizontally; they are referred to as *plagiotropic*.

According to these simple morphological characteristics, F. Halle has grouped all the known plants into less than 30 architectural models. Each of these models (e.g., Figure 6.2a-e) corresponds to a partic-

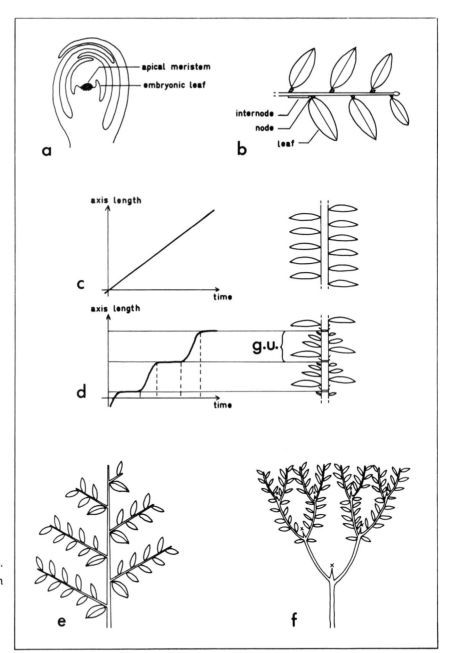

Figure 6.1
The apical meristem (a) terminates the leafy axis (b). (c) Continuous growth of an axis. (d) Rhythmic growth of an axis. (e) Monopodial branching. (f) Sympodial branching.

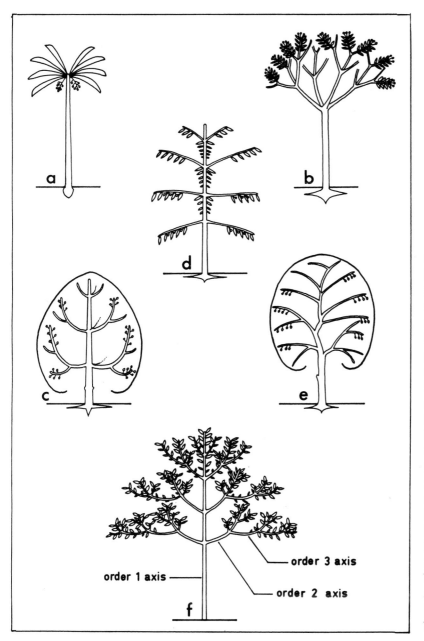

Figure 6.2

Five architectural models. (a) Corner's model: the single vertical axis (orthotropic) with spiral phyllotaxis bears lateral inflorescences (e.g., the coconut palm). (b) Leeuwenberg's model: the plant is constructed by a succession of "modules" derived from each other by three-dimensional sympodial branching (e.g., the common oleander). (c) Rauh's model: All the axes are orthotropic with rhythmic growth and branching; sexuality is lateral (e.g., pines). (d) Massart's model: The trunk is orthotropic and bears tiers of horizontal branches (e.g., fir). (e) Troll's model: The plant is built up by the superposition of axes which are all plagiotropic (e.g., elm). (f) The notion of branching order.

ular combination of the different possible types of axes. This notion does not correspond to the classic notion of habit because several models can exhibit a similar shape even though they function differently.

Each branch can on a plant or tree can be classified by a number called an *order number*. The hierarchy existing between axes clearly illustrates the notion of branching order (Figure 6.2f) in monopodial trees, studied in particular by C. Edelin [10]. The trunk is, by definition, an order 1 axis, the axes it bears are order 2 axes, and so on. The different axes of a given tree have specific morphological features and can thus be grouped into categories. The description of all the categories of axes of a tree belonging to a given architectural model represents its specific architectural unit [11]. These categories of axes are often connected with the notion of a branching order.

The architectural unit of a tree does not usually allow it to grow past a certain stage. After the tree has reached a certain size, it may duplicate its architectural unit according to a specific inherent strategy, referred to as "reiteration" by R.A.A. Oldeman [12]. A tall tree is thus a stack of reiterations, each of them representing a repetition of the architectural unit (Figure 6.3).

F. Halle and his co-researchers rationalized tree architecture by providing several fundamental concepts. The tree is analyzed and drawn by the botanist in a specific way so that the notions of reiteration and branching order appear explicitly. However, this approach is insufficient for quantitative analysis. In order to be clear, simplifications are used which are incompatible with the principle of mathematical modeling, based on the accuracy of its measurements. Furthermore, the notion of architectural model was first conceived for tropical plants for which the notion of architectural model is something caricatural (i.e., in *Terminalia*). It was based on the easy recognition of the branching order. In temperate areas, there are not as many architectural models, and they are less apparent.

Physiologists have their own approach, which is different but complementary. For them, what counts is not where the buds and axes are located along the tree trunk, but the tree's degree of differentiation, which we will refer to as its *physiological age*. P. Rivals [13] did a thorough study of the different categories of buds, their functional features and the axes they produce. He classified the axes according to their degrees of vigorousness as follows:

- The auxiblasts (Figure 6.4a) are the physiologically young vigorous axes (numerous long internodes $\{> 40\}$, essentially vegetative nodes).

Figure 6.3
Diagrammatic representation of
the developmental sequence of
a forest tree. (a and b) The "tree
of the future" expresses its
architectural unit. (c) The "tree
of the present" has developed a
large crown by means of
reiteration. (d) The "tree of the
past" is marked by the death
and by the gradual dislocation
of the crown; the limbs start
breaking, leaving stumps on the
trunk on which populations of
epiphytes become established in
tropical trees.

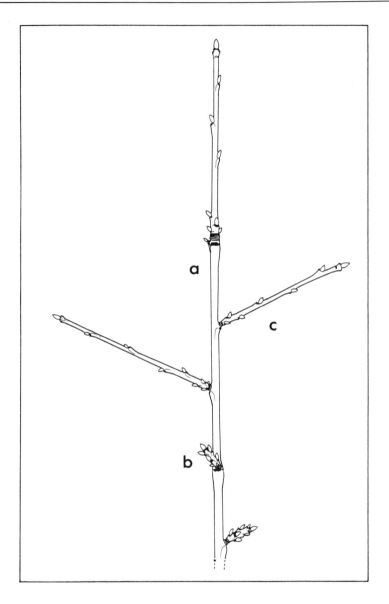

Figure 6.4
Branching system of *Prunus avium* **L.** (Rosaceae) illustrating the terms auxiblast (a), brachyblast (b), and mesoblast (c).

- The brachyblasts (Figure 6.4b) are the physiologically old short axes (a small number of short internodes {< 12}; they are floriferous).

Between these two poles, there are many intermediate types of axes referred to as *mesoblasts* (Figure 6.4c).

We should point out that there are several ways for auxiblasts to evolve into brachyblasts:

- By evolving along a vegetative axis. From the basis to the top of the trunk the vigorous growth units will progressively become a short G.U. (brachyblast), by "deflection."

- By successive branching. The higher the branching order of a branch is, the shorter its G.U.s tend to be.

- Along a growth unit, there can be an apical dominance which signifies that as we go down from the top of the G.U., the axillary branches tend to be less vigorous (acrotony phenomenon). Thus, there may be a whole series of axillary auxiblast branches in a single G.U. which are progressively becoming brachyblasts at the same time (for example, in a poplar tree). Because brachyblasts are first to undergo self-pruning, through time the notion of categories of axes becomes connected with a branching order.

Therefore, an order 1 axis can progressively take on the characteristics of an order 2 axis, and then those of an order 3 axis, and so on. Theoretically, a plant can locally bear axes of the same physiological age (degree of differentiation) or older than the axis of the bearing node. Reversions are possible when the plant undergoes a trauma (pruning, etc.).

The dialogue established between botanists and agronomists at the CIRAD Modeling Laboratory has allowed us to link the global point of view of the botanist with the local point of view of the physiologist in front of a tree.

6.3 STOCHASTIC PROCESSES FOR MODELING FUNCTIONAL FEATURES OF MERISTEMS

6.3.1 Qualitative Description of Meristem Structures and Functional Features

Quantitative modeling can only be based on figures characterizing measurable events. Qualitative analysis, described earlier, allows for the axes of a tree to be grouped into homogeneous populations, providing an essential starting point.

In our research, the basic event in the plant growth process we are concerned with is the appearance of a new internode at the tip of the leafy axis and its associated node which bears the axillary productions such as buds and inflorescences. (The term "inflorescence" usually refers to the mode in which the flowers of a plant are arranged in relation to the axis and to each other. It also can refer to the collective flower or blossom of a plant.) This new node event results from the functioning of the terminal meristem, and it characterizes the growth process. The axillary buds formed in the leaf axil can remain dormant for certain period of time. When they develop into an "axis" we call this step the "branching process." During the final phase, the meristems die and the axes undergo self-pruning. This is the death process. Growth, branching, and death are the three processes which affect the life of meristems.

In order to model their functional patterns, their qualitative formation, described in detail by Rivals [13], must be observed.

Minute embryonic leaves, separated by very short internodes, are produced in the buds by the terminal meristem according to a certain rhythmicity (apical growth). These productions, produced in the bud by organogenesis, are referred to as the *preformed part* (Figure 6.5a).

The production of new elements is invisible to the observer. Only the elongation and the unfolding of the reserve of preformed organs (internodes and leaves) can be seen (internodal growth) (Figure 6.5a). This elongation may affect some or all of the available preformed organs. Elongation is, of course, limited by organogenesis. These two types of functioning are not necessarily synchronized.

In temperate trees, there is generally a springtime elongation of the preformed part contained in the bud. After a resting period, of variable duration according to the tree's physiological age, the branch can start growing again to produce a neoformed part, which results for both apical and internodal growth. The auxiblasts function with neoformation, whereas brachyblasts do not have any. In late summer the branch stops growing and the bud dies or becomes dormant. It may then be subject to only apical growth. The growth unit (Figure 6.5b) is therefore the part of the axis located between two successive growth cessation points. The relations between two successive growth units differ from those between two consecutive internodes. The lapse of time between the appearance of two internodes can be short; several internodes often lengthen at the same time, whereas the lapse of time between the appearance of two growth units is longer (sometimes up

Figure 6.5
(a) Duration of observation. (b) Description of the meristem functioning.

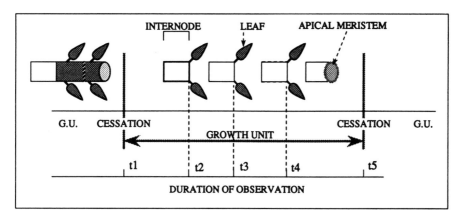

(a)

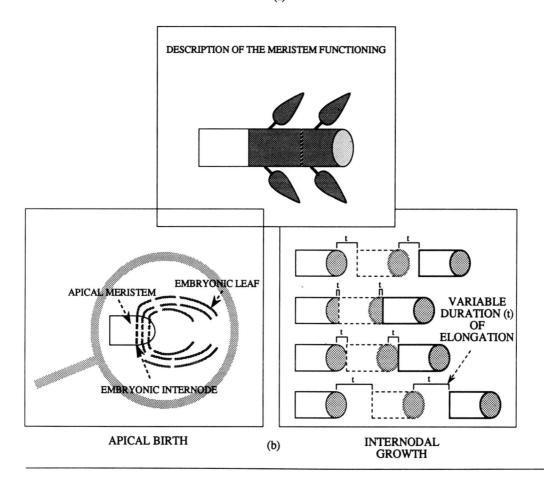

(b)

to 1 year). Therefore, the growth of the leafy axis depends on two interconnected processes:

1. The elongation of the internodes which form a growth unit.

2. The succession of the G.U.s which form a branch.

6.3.2 Random Characteristic of Meristem Functioning

This morphological description of terminal meristems obviously brings to mind queuing theory (W. Feller [14]), a form of probability theory useful in studying delays or lineups at service points. Apical growth is, by analogy, the law of customers' input (the stock of internodes produced, those who are waiting in line to be served), and elongation is the system's law of customers' output.

In practice, only the output law can be observed. Experimentally, this law appears to be random. Therefore, if the internodal growth of a population of leafy axes measured in internodes (belonging to the same clone of the same age) is observed during a given lapse of time, we notice that the latter are distributed according to characteristic bell-shaped curves. Experimentally, no correlation has been observed between successive growth periods for the species studied. It is impossible to predict the number of internodes this meristem will lengthen ahead of time.

The stochastic approach is therefore logical for this phenomenon. Although a specific event is not predictable, it is at least possible to control the statistical evolution of the population of buds with only a few parameters.

Another fundamental aspect of our modeling is the way time is managed in biological events. Plants grow during favorable periods. This means that the growth process between axes is structured and hierarchized. The growth rate is subject to variations, but the topological structure of the realized plant is not. Therefore, what counts for the structuring is the relative probability of the events and not how fast they appear. This concept is often used in agronomy – for example, when the sum of the temperatures necessary for an annual plant to flower is calculated.

It is also important to point out that plants are highly dependent on their environment. The development of the aerial structure depends on the soil, the root system, and atmospheric interactions with the canopy and the environment. Therefore, the parameters measured are essentially linked with exogenous conditions. This phenomenon makes it possible to compare the functional pattern of a given plant under different cultivation conditions so as to optimize the environment.

6.4 THE MATHEMATICAL MODEL AND ITS VALIDATION

If the stock of preformed internodes available in the meristem is sufficient, then a statistical process called the *Poisson process* is the simplest way to characterize the production of new internodes.

The Poisson process supposes that the probability for a new internode to appear during a given lapse of time is constant and that this appearance does not depend on the preceding appearances. Therefore, the distribution of internodes which appear during a given period follows the Poisson law, and the lapse of time between two appearances is exponentially distributed (Figure 6.6).

This distribution was experimentally discovered for the number of internodes produced per G.U. in the litchi tree [15] (see curve 1, Figure 6.6). This tropical plant grows by flushes, and always has a sufficient number of preformed internodes. Thus, the conditions for the Poisson distribution to function are satisfied. This distribution was also discovered for the number of internodes produced in some varieties of cultivated flax, and at the beginning of the elongation of the preformed parts of temperate trees per time unit (wild cherry, apricot).

If $P_n(t)$ is the probability for n events to be observed during the lapse of time t, and if λdt is the probability for an internode to appear during time dt, we obtain the following transition equations:

$$dP_0(t)/dt = -\lambda P_o(t) \tag{6.1}$$

$$dP_n(t)/dt = -\lambda P_n(t) + \lambda P_{n-1}(t) \quad \text{for } n \in N \tag{6.2}$$

The integration gives the probabilities at the moment t and gives the Poisson law:

$$P_n(t) = e^{-\lambda t}\frac{(\lambda t)^n}{n!} \tag{6.3}$$

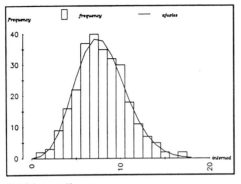

litchi (curve 1)

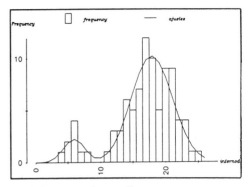

wild cherry tree (curve 3)

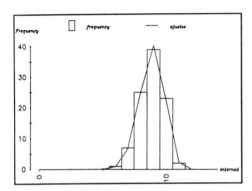

youg cocoa tree (curve 5)

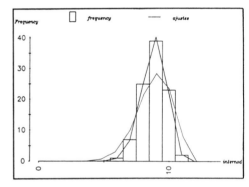

The biniomal distribution does not fit to the Solignac distribution.

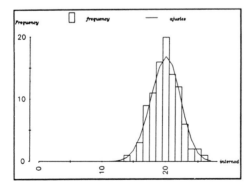

old cocoa tree (curve 6)

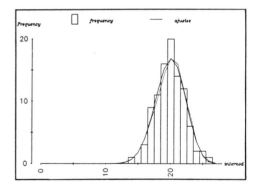

The binomial ditribution fits to the Solignac distribution.

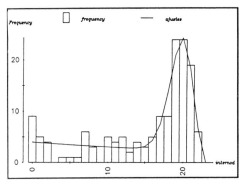

bamboo (curve 7)

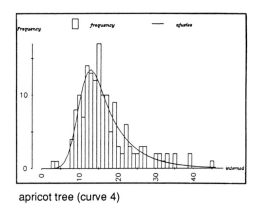

apricot tree (curve 4)

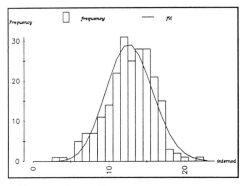

rubber tree (curve 2)

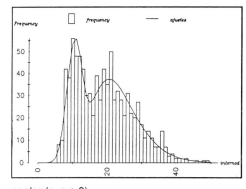

poplar (curve 8)

Figure 6.6

(Both pages.) Distribution functions for various plants (as designated in the figure).

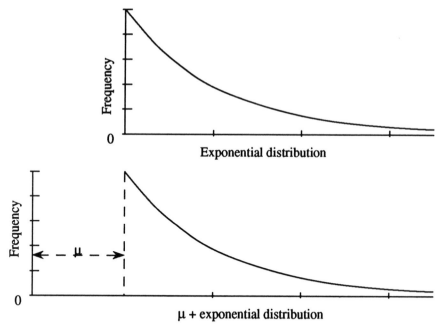

Figure 6.7
Mathematical approximation for
distribution functions.

6.4.1 Breakdown of the Poisson Process

When there is a sufficient stock of preformed internodes, two successive internodes can appear almost simultaneously. When this is not the case, a minimum time interval is necessary for the internode to be produced and elongated. Therefore the simplest hypothesis is that the delay between the elongation of two successive internodes is distributed according to the sum of a constant μ and the exponential distribution of parameter (Figure 6.7). If $\mu = 0$, we find the Poisson process again. If $\mu > 0$, we obtain a distribution with mean smaller than the variance (they are equal in the Poisson process).

In our laboratory, E. Elguero and co-workers [16] found the probabilities for this distribution, named the "Solignac distribution":

$$
\begin{aligned}
P(N(\tau) = k) &= e^{-\lambda(\tau - (k+1)\mu)} \sum_{i=0}^{k} \frac{\left[\lambda(\tau - (k+1)\mu)\right]^{i}}{i!} \\
&\quad -e^{-\lambda(\tau - k\mu)} \sum_{i=0}^{k-1} \frac{\left[\lambda(\tau - k\mu)\right]^{i}}{i!}
\end{aligned}
\tag{6.4}
$$

Using the central limit theorem (W. Feller [14]), this distribution will tend rapidly toward a normal distribution (as in the case of the Poisson process). Therefore, the approximation with the binomial distribution which also converges rapidly to the normal distribution becomes excellent. It functions as if the continuous time observed in N development stages, where an internode has probability b to lengthen, had been discretized. Thus, the growth process becomes purely binomial with the probability

$$P(x = k) = C_N^k b^k (1 - b)^N \tag{6.5}$$

It is shown that

$$N \approx \frac{\lambda\mu + 1}{\lambda\mu + 2} \cdot \frac{t}{\mu} \tag{6.6}$$

$$b \approx 1 - (1 + \lambda\mu)^{-2} \tag{6.7}$$

where λ, μ, t are the parameters of the above Poisson breakdown process. N corresponds to the number of fictive growth tests of the meristem called the "dimension." The associated probability b is called "meristem activity," and $(1 - b)$ is called the "rest probability."

The differences between the binomial distribution and the theoretical distribution of the breakdown of the Poisson process were observed for the first time when studying the growth of young cocoa trees (see curves 5 and 6 of Figure 6.6) over short periods of time. For longer lapses of time, the binomial distribution becomes quite acceptable.

The binomial process can be observed in many cases – for example, in the coffee tree [17], the cotton tree [18], bamboo [19] (see curve 7 of Figure 6.6), and the rubber tree [20] (see curve 2 of Figure 6.6) as well as in the neoformed parts of temperate tree leaf axes (poplar [21], wild cherry [22,23], and apricot [24,25]). In the case of neoformation, the apical growth slows down the internodal growth by limiting the number of internodes available for elongation. The breakdown of the Poisson process then appears quite logically.

It should be pointed out that in all cases the discretization of time in regular intervals, characteristic of binomial growth, approximates the continuous growth process very well. Moreover, it considerably simplifies the simulation methods (the Poisson distribution can be approached by the binomial distribution).

6.4.2 Interaction Between the Growth Process and the Cessation Process

When the growth process of a population of homogeneous leafy axes is interrupted on the same date, the internodal distribution can be observed directly (Poisson, binomial).

If the growth cessation periods are disposed according to a distribution, we will obtain a mixture arising from the combination of subpopulations interrupted at each cessation.

The final distribution has a probability

$$P(X = x) = \sum_{n=x}^{\infty} P(N = n) C_n^x b^x (1 - b)^{n-x} \qquad (6.8)$$

A very general case of the distribution of growth cessation periods is provided by the generative function $g(s) = (1 - w + ws)^{1/R}$, which can represent a whole family of discrete laws according to the values of w and R (binomial, negative binomial, geometrical, Poisson).

The combination of this cessation law with the growth process is expressed as follows:

$$
\begin{aligned}
g(s) &= \left[1 - w + w(1 - b + bs)\right]^{1/R} & (6.9) \\
&= \left[1 - wb + wbs\right]^{1/R} & (6.10)
\end{aligned}
$$

Therefore, the resulting function is much like the cessation law. This result is very important because it allows us to trace back to the cessation process from the final internodal distribution. Thus, the tedious monitoring of the successive stages of development is no longer necessary. With only one data survey at the final stage, the functional features of the G.U. can be deduced.

6.4.3 Experimental Observation of the Number of Internodes per G.U.

When the growth units from buds of the same physiological age are compared, we find that there are two main types of G.U.s in the trees examined so far:

1. G.U.s resulting from a single growth period of the meristems. The growth periods are all of the same duration. The final distribu-

tion of the number of internodes therefore results in a Poisson or binomial distribution (litchi tree, rubber tree, bamboo, etc.) – see curves 1, 2, and 7 of Figure 6.6;

2. G.U.s resulting from several growth cessation periods during development. Several cases occur which illustrate the meristem's aptitude to form a neoformed part after elongation of the preformed part during the springtime growth period. When accomplished, the opening up of the preformed part generally results in a binomial distribution.

Some of the meristems continue functioning as neoformations. Cessation of the neoformation in a tree can be almost simultaneous (wild cherry tree: binomial neoformation), staggered (poplar tree: negative binomial neoformation), or at death (Japanese elm tree, apricot tree: geometrical neoformation; in this case, the cessation probability is constant in neoformation – see curves 3, 4, and 8 of Figure 6.6).

G.U. Evolution in Tree Architecture

Along a tree's natural botanical gradients (deflection, branching order, acrotony on G.U.), the neoformed part decreases as the preformed part increases. At the same time, the internodes get shorter. This phenomenon was measured in particular on the keaki tree [26], the cherry tree [22,23], and the poplar tree [21].

In the trees that have been studied, a progressive evolution of the parameters of the internodal distribution law is observed according to the physiological age of the G.U.s. Tree uniformity is due to the fact that this law applies to the entire tree, characterizing its functional pattern.

Appearance Law for New G.U.s

New G.U. appearance is similar to internode elongation but at a different scale. In tropical plants (litchi, rubber tree), the process from G.U. to G.U. is binomial. The coffee tree is an exception because its G.U. has only one internode. Binomial growth was studied for the first time with this plant [16].

The time interval between two successive G.U.s is variable: 15 days for the coffee tree and up to several months for the litchi tree and rubber tree. For temperate plants, except for the juvenile polycyclisms, it is one year long.

The G.U.s are the visible marks of the tree's growth periods. In general, the separation between two G.U.s is marked by scars which remain visible for several years. On the scale of one tree, it is preferable to manage time according to the appearance of G.U.s.

The "dimension" refers to the number of formation tests on a new G.U.,and the "activity" refers to the success probability. The functional parameters of a tree can be estimated by the "crown method," the "mean variance link" [16], and so on, which we have developed to calculate the growth process. In order to analyze the mortality, the number of survivors of a population at a given phase are analyzed to determine the meristem viability.

6.5   THE BRANCHING PROCESS

When the terminal meristem produces new internodes, the latter bear nodes with axillary buds at the leafy axes. Branching can be instantaneous (referred to as *sylleptic branching*), or delayed (referred to as *proleptic*). Thus, there is a budding probability for dormant buds in terms of time.

With sylleptic branching, the branching test is done only once. If it is not positive, it will never bud. With proleptic branching, the budding probability is tested at each dimension unit.

Number of axilliary buds	2	1	0	Coupling
Sample	96	15	39	0.75

Links or "coupling" are often observed between axillary meristems of the same G.U. and must be taken into account. There are regions with branches, patches, and bare regions. Two examples can illustrate these links:

1. The coffee tree normally has two axillary buds per node. The absence of one or two axillaries should be randomly distributed. But, in fact, it tends to have either two axillaries or none.

2. The cecropia's trunk is formed with a series of similar internodes, and the branching process gives rise to tiers of branches inserted along the trunk. The number of branches per patch

and the interval between two tiers follow geometrical laws. The phenomenon is controlled by two probabilities:

    (a) The appearance law of a branch on a node (5% probability)

    (b) The tiers of branches formation law, if the first branch has appeared (70% probability for the branching process to continue)

This series of consecutive branches caused by coupling is very frequent (coffee tree, apricot tree, elm tree).

In well-individualized G.U.s, the branching can be located in privileged regions:

- The terminal region, called "acrotony" (poplar tree, apricot tree, etc.)

- The median region, called "mesotony" (rubber tree)

With acrotony, the physiological age of the axillary meristems increases as we go down the G.U. toward its base, changing from auxiblasts downward to brachyblasts. This fact is the sign of apical dominance. Branching may be affected by the orientation of the axillary meristems. For example, in the case of epitony, the axillary meristems which are oriented upward have a privileged budding. This fact complicates sampling.

To sum up, the branching process can be measured like the growth and death processes. However, it is the most difficult of the three to control in an architecture, owing to the obvious importance of conditional probabilities or coupling.

We will finish our account of this experimental mathematical model by insisting on the importance of the field work which must be carried out. The accuracy of the probabilities for the meristem functional pattern depends on the number of trees sampled. An unasked question cannot be answered, and there is always a possibility that an interesting particularity of a given tree may have been overlooked. When they render well, simulations are in this case a guarantee. The completed tool has shown that the static architectural measurements were sufficient to characterize growth processes. When in possession of growth follow-ups (coffee tree, cotton plant ...), we were actually in a position to test it. This is of great interest, since growth follow-ups are long and hard to implement. Therefore our method notably simplifies the experimentation.

6.6 SIMULATION OF TREE DEVELOPMENT

The simulation of tree development is the logical consequence of the research carried out on modeling. It is only possible if the main obstacles (managing time and physiological ages of meristems and tree space occupation) have been overcome. The problems which must be solved can be divided into five parts:

- The structuring of the data used to describe the functional pattern of a tree

- The stochastic growth engine which establishes the tree's topology

- The geometrical modules which can correctly position the vegetative elements spatially

- The important computer representation of the simulated tree resulting from the graphic database

- The creation of growth scenarios and animations

Our simulation project, called AMAP, has advanced along with our research. In 1979, the first version was designed by Ph. de Reffye for a Hewlett Packard computer, using HPL language, which controlled a plotter board. This version was designed for the coffee tree, but it can also be used to draw architectural models. Although this prototype was not well structured, it does contain the germs of the versions to come. The second version was designed in 1987 by M. Jaeger and Ph. de Reffye [27]. M. Jaeger [28] restructured the prototype program and designed a stack to manage the events. He separated topology and geometry, defined the notions of reiteration, organs, and structures, gave a typology of the qualitative features manipulated, and lastly introduced the "elastic line" notion. In this version, first written in Fortran and then in C, we come close to C. Edelin's [29] "architectural unit" notion, in which the concept of order is also essential. This version can accurately simulate young trees, but cannot handle the aging process. The third version, finished in 1990, is based on a notion of physiological age. The accuracy of this AMAP version, written in C by P. Dinouardd, M. Jaeger, and Ph. de Reffye in 1989, is satisfactory for arboriculturists' needs. It can simulate fruit trees (apricot, sweet cherry), temperate (elm, poplar) or tropical trees (rubber tree, litchi), and small plants (begonias, roses, orchids, flax, millet, etc.). It can simulate the different growth phases and follow the evolution of the meristems' functional laws.

6.6.1 Data Structure: "Reference Axis" Notion

The data are structured according to a "reference axis" [26] which translates the functional changes of the meristem as it gets older. This leads to a more flexible data organization. The notions of order, reiteration, organs, and structures are no longer needed. There are only meristems which function according to their physiological age. This version can produce accurate tree simulation from their juvenile phase to their adult phase. The functional parameters of the G.U.s are input with a data-capture program which automatically interpolates the parameters between two physiological ages. For example, the Japanese Keaki (*Zelkovva serrata*) was measured at the following stages: 1 year, 2 years, 5 years, 10 years, 25 years, 100 years. The intermediate evolution of the tree can be calculated by interpolating the description of these six stages.

6.6.2 The Stochastic Growth Engine

The growth engine, an expression used by J. Francon, Director of the Computer Laboratory at the University of Strasbourg, deals with the events which actually come about in a plant's life, in terms of time and probability. Using the prefixed order (Figure 6.8), the evolution of all the axillary buds can be explored one by one. The Monte Carlo method, based on the generation of random numbers, is used to reproduce the meristem lengthenings, cessations, and ramifications according to their probability laws. The G.U.s of a given physiological age receive their laws from the reference axis. The axillary buds' physiological age differences vary according to the acrotony. The meristems evolve step by step during their growth period.

Time is discretized, becoming the tree's "dimension." Each tree is controlled by a clock. At each signal, the functional state of all the meristems is tested in terms of their probabilities. Rhythms may differ. Some meristems can function quicker or slower. The dead parts are automatically pruned. The death of a meristem can be due to an accident or natural aging. Once all the buds have been tested during their growth period and in terms of their physiological ages, a topological structure can be established which describes the number of internodes, leaves, flowers, branches, and so on, without having to draw the tree (Figure 6.9).

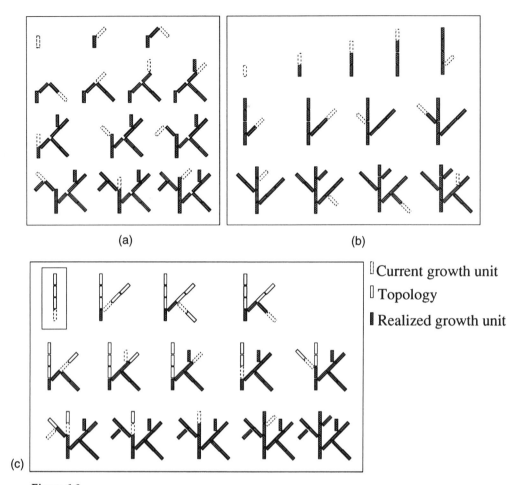

(a) (b)

(c)

Current growth unit

Topology

Realized growth unit

Figure 6.8
(a) Prefix order: In the prefix order simulation, only branching nodes are to be stored. (b) Order by order: In the order-by-order simulation, each order is computed totally axis by axis. All nodes of the previous orders are to be stored. (c) Data structures used in a map: Our simulation program computes the nodes in prefix order, but the topology (activity of the bud) is computed only for an axis and for the whole axis.

A parallelized version of this program is being developed by F. Blaise [30] at our laboratory. It will enable the control of exogenous events which modify the tree's functional pattern (temperature, ...) using a schedule system. The meristems can also interact with one another. This allows for an interesting approach to the disturbances which arise

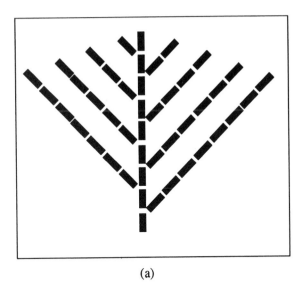

(a)

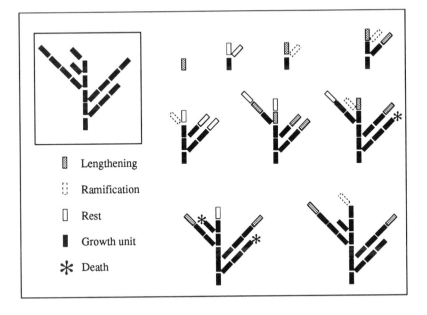

(b)

Figure 6.9
(a) The whole possible tree realized if all the clock cycles give birth to an internode. (b) The realization, according to a succession of stages, of different growth tests on the apical and axillary meristems. In this example, it is considered that each signal of the clock is annual (the G.U. is restricted to a single internode). We note a first lengthening at 1 year, two rests at 2 years, and the two lengthenings, one of which is making a ramification at 3 years (since it is a new axis). The growth of this tree ends at 9 years.

Figure 6.10
Parallel growth generation allows response of the tree growth according to light and space. (a) Light constraint on growth. (b) Space constraint on growth. See insert for color representation.

(a) Light constraint on growth

(b) Space constraint on growth

among neighboring trees and among buds on the same tree. This also allows simulation, with light constraints and space constraints (Figure 6.10). The prefixed tree exploration method is not appropriate for this approach (Figure 6.11).

6.6.3   The Geometrical Construction of Trees

The length of internodes and their evolution, thickening, branching and phyllotaxis angles, tropisms, and wood elasticity are all parameters necessary for the three-dimensional visualization of topological trees simulated according to field data. With this data measured on the tree and keyed onto the reference axis, the program can then properly locate and orient the topological events, a new internode, leaves, and so on. A considerable number of geometrical operations are needed in order to form a big tree. The introduction of geometry allows the volume of wood produced and the foliar surface to be measured and the yield estimated, considering all other factors are equal.

6.6.4   Structure of the Simulation-Based Graphic Data Base: The "Elastic Line"

The simulated topological elements are stored in an "elastic line" file designed by M. Jaeger [28]. This elastic line contains all the informa-

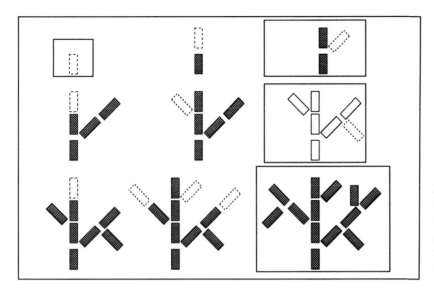

Figure 6.11
In this new growth engine, the tree is built step by step like in nature. It needs to store every node. Steps 1, 2, 6, and 9 are realistic steps.

tion necessary for the representation of tree organs: transformation matrix, age of organs, symbol number, and so on. It can be manipulated on a high level. It allows for classical sorting and geometrical operations. The tree can be visualized by linking the elastic line and a three-dimensional base of organs (leaves, internodes) on the display (Figure 6.12).

6.6.5 Producing Natural Scenarios and Simulating Plant Growth

In order to simulate a field of plants, a park, or a garden, it is important to have software capable of producing three-dimensional scenarios where the plants can be interactively positioned. Furthermore, with computer graphics, scenarios and animations can be used to check the results of the agronomists' and landscape architects' simulations. From a computer graphics standpoint, natural landscapes can be comprised of thousands of polygons. Because the degree of fineness for the visual rendering cannot be calculated by classical algorithms (shadow, etc.), M. Jaeger developed some algorithms specially designed for these complex scenarios and P. Dinouard dealt with the animation of these scenarios (changes of viewpoint, plant growth) (Figure 6.13).

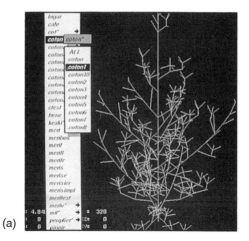

(a)

The line tree:
— result of the simulation (cotton tree)

Pattern:
— a leaf.

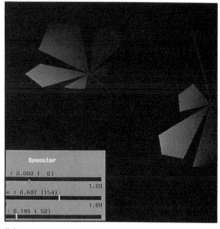

(b)

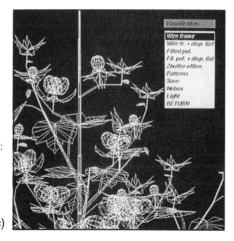

The line tree with its pattern:
— tne cotton tree in wire frame.

(c)

The line tree with its pattern:
— illumination with Gouraud shading on the cotton tree.

(d)

Figure 6.12
(a) The line tree – result of a simulation (cotton tree). (b) Pattern – a leaf. (c) The line tree with its pattern – the cotton tree in wire frame. (d) The line tree with its pattern – illumination with Gouraud shading on the cotton tree. See insert for color representation.

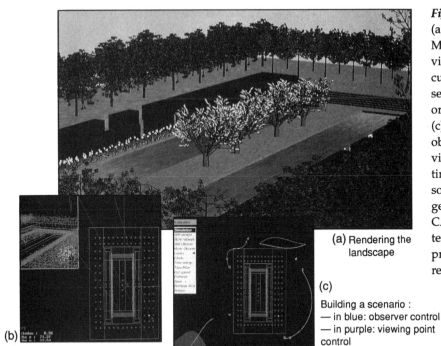

(a) Rendering the landscape

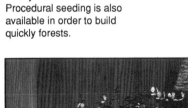

(b)

Making the landscape
— in red: viewing pyramid
— in yellow: the current tree
Procedural seeding is also
available in order to build
quickly forests.

(c)

Building a scenario :
— in blue: observer control
— in purple: viewing point
control
— in yellow: time control

Figure 6.13
(a) Rendering the landscape. (b)
Making the landscape (in red:
viewing pyramid; in yellow: the
current tree). Procedural
seeding is also available in
order to build quickly forests.
(c) Building a scenario. In blue:
observer control; in purple:
viewing point control; in yellow:
time control. (d) The redering
software can use terrain
generators and well known
CAD files. (e) Rendering with
texture mapping and shadow
processing. See insert for color
representation.

The rendering software can use terrain
generators and well known CAD files.

(d)

Rendering with texture
mapping and shadow
processing.   (e)

173

6.7   TREES: FROM THE FIELD TO THE COMPUTER

In practice, several successive stages are necessary for the correct simulation of a given tree.

1. Botanical field observations (Figure 6.14a and Figure 6.15a) provide precise qualitative data concerning the architecture of a plant and its growth pattern. These observations include precise drawings which clearly show the main events characterizing the plant's development (establishment of the architectural unit, reiteration strategy, aging).

2. The plant is then numerically measured. The borne axes are identified and described from the top downward according to their location along the bearing axes. Observations are taken through time. This phase is tedious and lasts several months. Because of the agronomical vocation of the CIRAD, it is usually done on plants cultivated in fields. The trees measured must be from homogeneous populations, preferably clones (e.g., poplar trees, coffee trees). The distributions of the number of internodes per G.U. are recorded, as well as the locations of the branches and the deaths of meristems. Geometrical measurements complete the data (length, angles of insertion, diameter, etc.). A database may then be created and interrogated to obtain statistics for events corresponding to a physiological age determined by the meristems. Numerical algorithms are used to estimate the G.U.s functional patterns (Figure 6.14b). These were specially designed for tree measurements by our Laboratory's mathematician (E. Elguero, 1989). It can take up to several months to exploit the database based on field observations.

3. The third phase is the creation of a reference axis which can take into account the different laws obtained from the trees, as well as their architectural evolutions. Theoretically, this phase should only take a few days.

4. The last phase is tree simulation (Figure 6.15b) on a graphics workstation. If the tree has been analyzed correctly, the simulations should be very similar to the trees studied. A parameter file is then established which can accurately reproduce how this tree will develop in its environment.

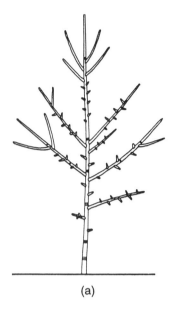

(a)

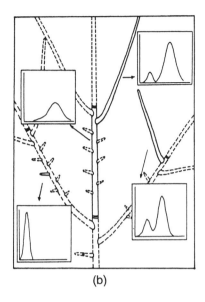

(b)

Figure 6.14
Wild cherry tree – *Prunus avium* L. (Rosaceae). (a) Botanical field drawing, using architectural concept. (b) Statistical evolution of internode per growth unit according to tree architecture.

(a)

(b)

Figure 6.15
Wild cherry tree – *Prunus avium* L. (Rosaceae). (a) Photography, (b) Tree simulation. See insert for color representation.

6.8 SCIENTIFIC FIELD OF APPLICATION

Plant modeling methods are of interest in many areas: teaching, agronomy, landscape architecture, and computer graphics. Teaching plant development and architecture could be made very attractive and educational if quick-motion animated films were used to illustrate the strategy a tree uses to occupy space. Films of this type have been made by our laboratory using stop-motion films of plant growth simulations (coffee tree, wild cherry tree, Japanese elm tree, litchi tree, etc.). The visualization of the concepts defined by F. Halle (architectural models, reiteration) helps to clarify them.

The quantitative aspect of this modeling makes it particularly easy for applications. The plant measured is expressed locally by statistical distributions which characterize the random functions of the events actually realized (formation of internodes, branches, etc.). These random functions express the interaction between the plant and its environment. If, for example, the environment is modified (in controlled irrigation density agronomical experiments), there can be direct and accurate appreciation of its influence on the plant's development. Agronomical optimization thus becomes possible by interpolating the different controlled situations to obtain the optimum.

Other types of applications are also possible: identification of yield factors (number of flowers produced, foliar surface, wood volume), studies of the genetic variability of a species' architecture, and studies of the light interception of leaves.

Our Modeling Laboratory has studied industrial tropical plants (coffee tree, rubber tree, cotton tree, cocoa tree, litchi tree, cassava, millet) and temperate plants (poplar tree, apricot tree, peach tree, wild cherry tree, elm tree, begonia, flax) in collaboration with the French professional agronomical institutes (INRA,[1] ORSTOM,[2] CTIFL,[3] IDF,[4] ITL,[5] and others).

In the industrial field of computer aided design(CAD), this is obviously an interesting tool for the composition of natural landscapes with seasonal evolution. With the AMAP program designed by our Laboratory, plants can also be selected for landscape architects. Parks, gardens, and other urban open spaces can be designed and their de-

---

[1] Inst. National de la Recherche Agronomique.
[2] Inst. Francais de Recherche Scientifique pour le Dév. en Cooperation.
[3] Centre Technique Interprofessionnel des Fruits et Legumes.
[4] Inst. pour le Développement Forestier.
[5] Inst. Technique Agricole du Lin.

velopment predicted. This tool is quite helpful for landscape architects because it allows them to find the best solutions and avoid mistakes. The AMAP program is also equipped with a library of commonly used plants which is progressively updated (trees, shrubs, flowers, grasses, etc.). It is easy to integrate in the existing CAD program.

Lastly, because AMAP is easy to use it has brought plants, real or imaginary, into the world of computer graphics, providing the latter with live procedural objects and contributing to the set of objects modelers can use for scenario design.

ACKNOWLEDGMENTS

The figures in this chapter have been produced by D. Barthelemy, Y. Caraglio, E. Elguero, M. Jaeger, and M. Gerenton.

REFERENCES

[1] F. Halle and R.A.A. Oldeman, *Essais sur l'Architecture et la Dynamique de Croissance des Arbres Tropicaux*, Masson, Paris (1970).

[2] F. Halle, R.A.A. Oldeman, and P.B. Tomlison, *Tropical Trees and Forests: An Architectural Analysis*, Springer-Verlag, Berlin (1978).

[3] M. Aono and T.L. Kunii, Botanical Tree Image Generation, *IEEE Comput. Graphics Appl.* **4**(5), 10-33 (1984).

[4] A.R. Smith, Plants, Fractals and Formal Languages, *Comput. Graphics* **18**(3), 1-10 (1984).

[5] A. Lindenmayer, Paracladial Systems, in *Automata, Languages, Development*, A. Lindenmayer and G. Rozenberg (eds.), North-Holland Publishing Company, Amsterdam (1976).

[6] P.E. Oppenheimer, Real Time Design and Animation of Fractal Plants and Trees, *Comput. Graphics* **20**(4), 55-64 (1986).

[7] G. Eyrolles, J. Francon, and G. Viennot, *Combinatoire pour la Synthèse d'Images Réalistes de Plantes*, Actes du Deuxieme Colloque Image, CESTA, pp. 648-652 (1986).

[8] G. Gardner, Simulation of Natural Scenes Using Textured Quadric Surfaces, *Comput. Graphics* **18**(3) (1984).

[9] P. Prusinkiewicz et al., Development Models of Herbaceous Plants for Computer Imagery Purposes, *Comput. Graphics* **22**(4), 141-150 (1988).

[10] C. Edelin, *L'Architecture Monopodiale: l'Exemple de Quelques Arbres d'Asie Tropicale*, Thèse de Doctorat ès Sciences, Université des Sciences et des Techniques du Languedoc, Montpellier, France (1984).

[11] D. Barthelemy, C. Edelin, and F. Halle, Architectural Concepts for Tropical Trees, in *Tropical Forests: Botanical Dynamics, Speciation and Diversity*, L.B. Holm-Nielsen, I. Nielsen, and H. Balslev (eds.), Academic Press, London, pp. 89-100 (1989).

[12] R.A.A. Oldeman, *L'architecture de la forêt Guyanaise*, Mem. No. 73, O.R.S.T.O.M., Paris, p. 204 (1974).

[13] P. Rivals, Essai sur la Croissance des Arbres et sur Leurs Systèmes de Floraison, *Journ. Agric. Trop. Botan. Appl* **XII**(12), 655-686 (1965); **XII**(1-2-3), 91-122 (1966); **XIV** 67-102 (1967).

[14] W. Feller, *An Introduction to Probability Theory and its Applications*, Vol. 1, third edition, John Wiley & Sons, New York (1950).

[15] E. Costes, *Analyse Architecturale et Modelisation du Litchi*, Thèse de Doctorat, Université des Sciences et Techniques du Languedoc, Montpellier, France (1988).

[16] P. de Reffye, E. Elguero, and E. Costes, *Growth Units Construction in Trees: A Stochastic Approach*, 9th Seminar of the Theoretical Biology School, Solignac, (21-23 September 1989).

[17] P. de Reffye, *Modélisation de l'Architecture des Arbres par des Processus Stochastiques. Simulation Spatiale des Modèles Tropicaux Sous l'Effet de la Pesanteur. Application au Coffea Robusta*, Thèse de Doctorat ès Sciences, Université de Paris-Sud, Centre d'Orsay (1979).

[18] P. de Reffye et al., Modélisation Stochastique de la Croissance et de l'Architecture du Cotonnier. I. Tiges Principales et Branches Fructifères Primaires, *Coton Fibres Top.* **XLIII**(4) 269-291 (1988).

[19] P. de Reffye, P. Dinouard and P. Dabadie, *Modélisation de la Croissance et de l'Architecture d'un Bambou*, Laboratoire de Modélisation du CIRAD/GERDAT, Institut de Botanique de Montpellier, Unpublished internal report (1988).

[20] P. de Reffye et al., *Rapport Préliminaire sur la Modélisation de l'Architecture de 3 Clones d'Hévéa,* Convention IRCA/CIRAD, unpublished internal report (1989).

[21] Y. Caraglio et al., *Le Peuplier, Modélisation et Simulation de Son Architecture,* Convention IDF/CIRAD, unpublished internal report (1990).

[22] Y. Caraglio et al., *Le Merisier, Modélisation et Simulation de Son Architecture,* Convention IDF/CIRAD, unpublished internal report (1990).

[23] D. Fournier, *Modélisation de la Croissance et de l'Architecture du Merisier (Prunus avium L.),* Diplôme d'Agronomie Approfondie, Ecole Nationale Superieure Agronomique de Montpellier (1989).

[24] C. Chavaneau, *Modélisation de la Croissance et de l'Architecture de l'Abricotier,* Diplôme d'Agronomie Approfondie, Ecole Nationale Superieure Agronomiqe de Montpellier (1989).

[25] E. Costes et al., *Modélisation de l'Architecture Aérienne de l'Abricotier Prunus armeniaca L.,* Convention Laboratoire de Modélisation du CIRAD/GERDAT et le CTIFL, unpublished internal report (1989).

[26] P. de Reffye, P. Dinouard, and D. Barthéléy, *Architecture et Modélisation de l'Orme du Japon, Zelkova serrata (thunb.) Makino (Ulmaceae): La Notion d'Axe de Référence,* Actes du 2éme Colloque International sur l'Arbre, Montpellier, France, n° hors série A7, pp. 251-266 (Sept. 10-15, 1990).

[27] P. de Reffye et al., Plant Models Faithful to Botanical Structure and Development, *Comput. Graphics* **22**(4), 151-158 (1988).

[28] M. Jaeger, *Représentation et Simulation de Croissance des Végétaux,* Thèse de Doctorat, Université Louis Pasteur de Strasbourg (1987).

[29] C. Edelin, *Images de l'Architecture des Conifères,* Thèse de Doctorat 3ème Cycle, Biologie végétale, Université de Monpellier II, France.

[30] F. Blaise, *Simulation du Parallelisme dans la Croissance des Plantes et Applications,* Thèse de Doctorat, Université Louis Pasteur, Strasbourg (1991).

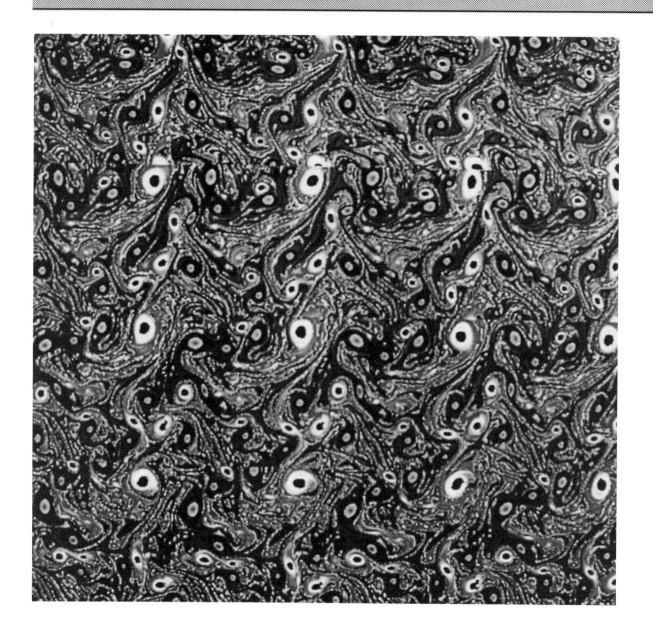

7

Scientific Display: A Means of Reconciling Artists and Scientists

Jean-Francois Colonna
CNET-LACTAMME
Ecole Polytechnique
91128 Palaiseau Cedex FRANCE

7.1 INTRODUCTION

We live in an age where there is increasing interplay between scientific and artistic disciplines. By the end of the decade almost all advances in science and art will rely partly on the computer and advanced technology. Moreover, humans will not be able to rely on any one single field of knowledge to make significant advances.

Since their rapid growth following the Second World War, computers have changed the way we perform scientific research, conduct business, create art, and spend our leisure time. The computer has also changed our perception of visual art.

From my own experience, as exhibited in the figures for this chapter, I have found that computer graphics is a powerful vehicle for artistic expression. The line between science and art has always been a fuzzy one; the two are fraternal philosophies formalized by ancient Greeks such as Eratosthenes and Ictinus. Today, computer graphics helps reunite these philosophies by providing scientific ways to represent natural, mathematical, and artistic objects. Computer graphics representations of the behavior of certain mathematical processes has revealed a surprising variety of beautiful, and sometimes unpredictable, patterns and surfaces. Many attractive patterns are graphical representation of intricate curves called fractals – which are described later in this chapter.

Science seeks to develop a coherent set of knowledge pertaining to certain categories of objects or phenomena. To comprehend our physical environment, we usually have two means at our disposal – observation and experiment. By observation I mean scrutiny through the use of our senses, particularly our visual senses. Unfortunately, as science progresses, direct observation of objects and phenomena of contemporary interest becomes more difficult as the objects and phenomena being studied become more abstract. No physicist, for example, has directly seen an "elementary" particle. Experimentation, which involves the notion of contact or manipulation of objects and phenomena, also becomes more difficult in other contemporary areas of interest. None of today's cosmologists, for example, are capable of manipulating the universe to explore its behavior in various situations. Increasingly, it is necessary to resort to numerical simulation of physical systems, drawing on a "virtual" system expressed through appropriate models to understand the behavior of corresponding physical systems.

Mathematics is a powerful language of description through which the laws of nature can be expressed. However, mathematics may also serve as a language of prediction, extending our understanding of observable objects and phenomena into the realm of nonobservable objects and phenomena (e.g., general relativity and the theories of elementary particles [1]). This potential "predictive" ability of mathematics is particularly important as a basis for the models used in numerical simulations of physical systems. Scientific visualization – the use of powerful computer graphics to gain insight into complicated data – allows researchers to visually study numerically simulated systems, drawing on the interpretive capabilities of our visual responses to better "understand" the behaviors hidden in the numerical data from the simulation.

However, the scientific method is not the only path towards a better understanding of our environment. The rules and techniques of art, for example, coupled with our emotional response to colors and shapes in the world around us, can be leveraged in the study of scientific data. These two paths seldom cross. There are very few major works of art influenced by the science of their day and, conversely, very few scientific theories that make use of the sensual harmonies contained within works of art.

The visual display of numerical results of scientific, numerical simulations, like any picture, will certainly trigger a variety of emotional, qualitative visual responses. Used properly, these visual responses can aid in understanding the behaviors of the artificial systems being simulated. However, used casually, they may provoke unexpected visual responses which distort the viewer's understanding of the results being displayed. Visualization used to present the results of a scientific experiment brings together the strict, rational approach of science and the aesthetic qualities of art. The reconciliation of these aesthetic and scientific objectives is the subject of this chapter.

7.2 THE NOTION OF VIRTUAL EXPERIMENTATION

Scientific experimentation normally involves using instrumentation (probably in a laboratory) to stimulate and then to measure the resulting behavior of physical objects or phenomena. Such "real" experiments contrast with "virtual" experimentation (a term we use in preference to "numerical" experimentation). The basis for a "virtual" experiment (assumed performed by a computer using numerical techniques) is a model of the physical or abstract system being simulated.

A real system is modeled by the various interactions within and among the various parts of the system, the model being provided by analytic equations. The mathematical models are combined, manipulated, and extended until predicted behavior matches observed behavior. The *formal approach* is to seek an analytic solution of the model equations for the physical system being studied. However, such analytic solutions are often impractical, or even perhaps impossible. Even simple systems governed by deterministic equations can display behavior which is difficult to capture in analytic solutions (e.g., chaotic behavior). The *numerical approach*, using the same underlying mathematical equations, proceeds by means of numerical solutions of those equations according to the representation of the system being studied. The results of such numerical solutions are often presented in the form of visual representations of the behavior of the system.

Virtual experimentation, therefore, assumes that the underlying analytic models of the simulation are correct (much as we presume that a real object is "correct") and explores the range of behaviors produced by that model (much as real experiments explore the range of behaviors of real objects). The wealth of numerical data developed during the numerical simulation is then massaged to provide a visual display of the results of the simulation. By viewing the results of this display, the researcher can adapt the parameters of the model or request different "views" of the numerical results. The researcher, the numerical calculations, and the displayed images form a feedback loop through which the complex behavior of the simulated system is explored.

7.3 PICTURE SYNTHESIS: A SCIENTIFIC TOOL

Vision is the most highly developed of our human senses for reception, isolation and understanding of information about our environment. Vision provides a global perception of colored shapes against a changing, moving, and noise-filled background. The idea of using the eye as the main tool in the analysis of numerical results is therefore quite natural.

The pictorial representation of experimental results (whether from virtual or more traditional forms of experimentation) can provide a global representation, rather than a point-by-point representation of the results, through the use of colorful forms and lines. This means that elements which are spatially distant from each other can be juxtaposed and connected by being presented in a similar fashion (e.g., in a given virtual experiment, all particles with a velocity between V and $V + \epsilon$ can be presented in the form of points of the same color). In this case, the viewer's visual responses can work in concert with the objective of efficiently presenting the vast number of numerical results. Complex forms will become distinct and the scientist may be able to ascertain a hidden order in the numerical results. As others have pointed out: *Scientific visualization is the art of making the unseen visible.*

Thus, pictures intended as little more than an efficient method for presenting a vast amount of numerical data can be a means of discovery, used to discern the unexpected or to observe something that no physical instrument can show – for example, the formation of a chromosome from the double DNA spiral or the dynamics of a collision between two galaxies. Like the various components of a musical orchestra which all come together to create a useful whole, the scientist, numerical calculation, and picture synthesis all work together to form a scientific instrument, as seen in the following examples.

An Instrument for Synthesis. Visualization is useful for displaying the results of numerical simulations. Two good examples are fractal landscapes (Figure 7.1) and fractal aggregates (Figure 7.2). The word "fractal" was coined in 1975 by mathematician Benoît Mandelbrot to describe an intricate-looking set of curves, many of which were never seen before the advent of computers with their ability to quickly perform massive calculations. Fractals often exhibit self-similarity, which means that various copies of an object can be found in the original object at smaller size scales. The detail continues for many magnifications – like an endless nesting of Russian dolls within dolls. Some of these shapes exist only in abstract geometric space, but others can be used as models for complex natural objects such as coastlines and blood vessel branching. Interestingly, fractals provide a useful framework for understanding chaotic processes and for performing image compression. The dazzling computer-generated images can be intoxicating, motivating students' interest in math more than any other mathematical discovery in the last century.

An Instrument for Validation. The results of numerical simulations and laboratory tests become easier to compare when the representative codes used are the same (see Figure 7.3).

An Instrument for Developmental Assistance. Generally speaking, a model is a complex object in terms of both its mathematical formulation and its expression in the form of a computer program. Numerous conceptual and technical errors can arise at various stages of development of a virtual experiment. Often, such errors become apparent in the form of spatial or temporal discontinuities which the eye will pick up immediately (see Figure 7.4). In addition, sensitivity to initial values or the precision of calculations can be more easily studied (see Figure 7.5).

An Instrument for Comprehending Abstract Concepts. Picture synthesis can provide a representation (in some cases, an arbitrary representation) of abstract concepts. This facility gives mathematics (in particular pure mathematics) the status as an experimental science, which it once enjoyed at the dawn of history (see Figure 7.6).

An Instrument to Manipulate Inaccessible or Invisible Objects. Picture synthesis also provides a means of seeing phenomena (when described by a valid model) that no other instrument can show either because they are too short-lived or long-lived (temporal dimension) or because they are too small or too large (spatial dimension) (see Figure 7.7 and 7.8). However, picture synthesis also provides a means of manipu-

Figure 7.1

(Opposite page) Fractal geometry, a mathematical concept explored and developed by Benoît Mandelbrot [4], can be used to model and describe numerous phenomena that were once viewed as unrelated. As a consequence of ideas developed in this field, fluctuations of the Stock Exchange, the advancement of forest fires, and the structure of borders between two media that are diffusing on top of each other can be described using the same language. Fractal geometry is used to model numerous natural phenomena and is, therefore, a powerful tool in the field of picture synthesis. This figure shows a set of completely "imaginary" landscapes calculated using equations for the geographical relief, the mist and clouds, and the dynamics of the clouds [5]. The artist can then make different use of these tools, transforming mathematics into a malleable matter, a sort of futuristic marble or clay – creating worlds that are either similar to our everyday surroundings or diametrically opposed to it. We then have all these worlds at our disposal in the form of virtual realities ready for exploration.

←

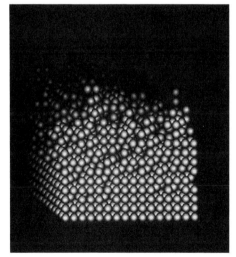

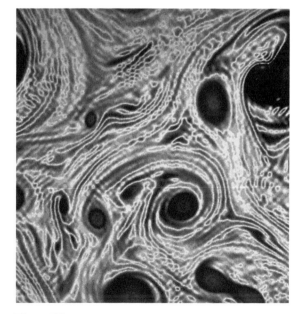

Figure 7.2

Representation of a three-dimensional fractal aggregate which can be used to model a porous body or the interface between two media diffusing on top of each other. This complicated three-dimensional object is difficult to visualize without three-dimensional computer graphics. See insert for color representation.

Figure 7.3

Display of a two-dimensional curl field. (Model: Marie Farge, LMD/ENS.) See insert for color representation.

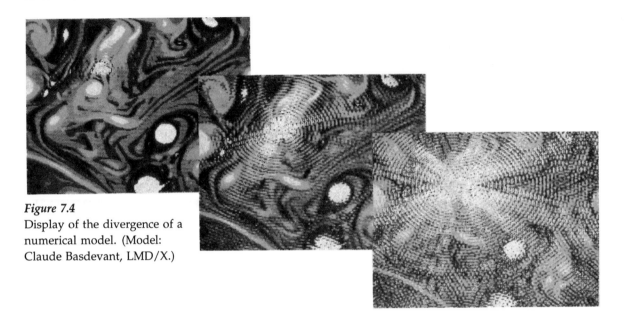

Figure 7.4
Display of the divergence of a
numerical model. (Model:
Claude Basdevant, LMD/X.)

Figure 7.5
Velhulst dynamics provides a simple model for increase in
animal populations. To compute this pattern, use the
formula $S_n = (1 + r) \cdot S_{n-1} - r \cdot S_{n-1}^2$, where r is the rate
of growth, S_0 is the initial population, and S_n is the
population after n units of time. This figure superposes the
results of computations preformed using representations of
numbers using 32 and 31 bits respectively. For 512 values of
r lying between 2.0 and 3.0 (represented on the y-axis), 32
values of S_n^{32} and S_n^{31} (on the abscissa, where exponents 32
and 31 represent the precision of the computations) are
obtained with n varying between 5000 and 5031. The same
initial value $s_0^{32} = S_0^{31} = 0.50000$ is used. Red color denotes
the use of 32 bits, green denotes 31 bits, and blue denotes
the inevitable "collisions" caused by the quantization of the
picture because it is not continuous. If fact, outside
nonchaotic areas, blue pixels are the exception and do not,
in general, correspond to identical values of n. What do
these differences tell us about the ability of computer to
usefully perform all types of calculations? See insert for
color representation.

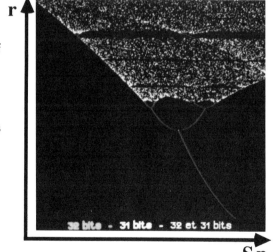

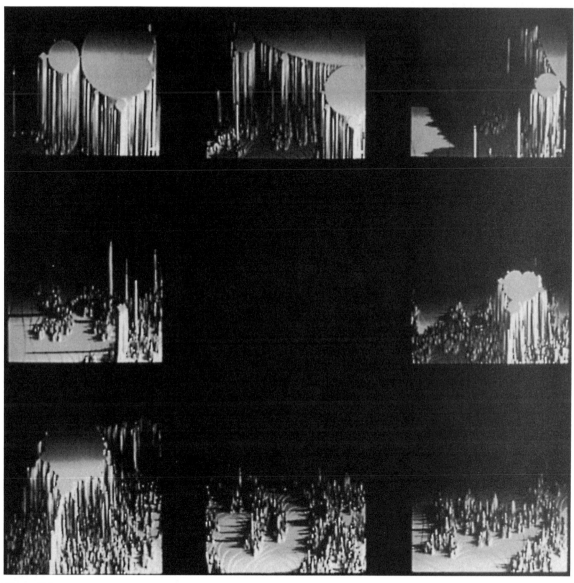

Figure 7.6

Magnification (by a factor 10^6) of a three-dimensional representation of a mathematical object called the Mandelbrot set $\mathcal{M}$. This intricate fractal object (i.e., an object with a structure that is repeated using every observation scale) is obtained using an iterative method. For each pixel C in the complex plane, the following series is defined: $Z_{n+1} = Z_n^2 + C$, where $Z_0 = 0$. During this iteration, initial points may explode to infinity or remain bounded. Those initial points that do not wander off to infinity during the course of the iteration comprise the Mandelbrot set.

Figure 7.7
Display of the aggregation of polymer strings on a macroparticle.

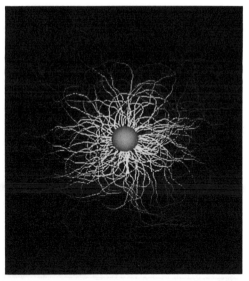

Figure 7.8
A proton. Protons and neutrons, once considered as elementary particles, are now described as composite objects. The "standard" model of particles and their interactions is based on bosons ("force vectors") and fermions ("matter"). The latter category contains the leptons and the quarks. This image shows a stereo view of a proton as defined in this model. It consists of three "real" quarks in the form of three sets of contiguous spheres that are, respectively, red, green, and blue, and a "sea" of virtual particles (gluons, quarks, and anti-quarks) produced by quantum fluctuations of the vacuum. (Color in this case is a notion similar to that of electric charge and is in no way related to what we would commonly call "color.") Statistically, the virtual pairs are concentrated between the real quarks according to color. The colored "small cylinders" represent gluons, the bosons that are responsible for the strong interaction (one of the four basic forces to which only quarks are sensitive). Thus, through the use of these new tools, all scales from galactic to subatomic are accessible to our eyes. See insert for color representation.

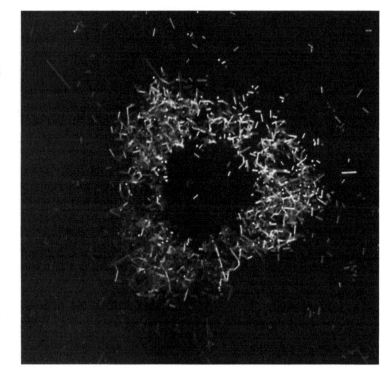

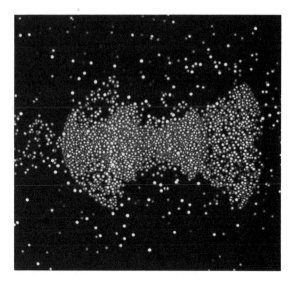

Figure 7.9
The collision of two galaxies. Only the stars in the smallest galaxy are represented. Characteristic structures known as "shells" are then formed and have since been recognized in photographs of the sky. (Model: Francoise Combes and Christophe Dupraz, Observatoire de Meudon.)

lating phenomena that are inaccessible in a laboratory. After having developed the corresponding model, a scientist can study, for example, the evolution of the earth's climate subsequent to a nuclear catastrophe, the collision of two galaxies (see Figure 7.9), or the "gravitational lens" effect in the vicinity of a black hole (see Figure 7.10).

An Instrument for Communications. For presentations of results to various audiences (professional, educational, and public), pictures are a vital means of communication. Computer graphics representations of results can be used to pinpoint details (see Figure 7.11), illustrate an abstract concept (see Figure 7.6) or assist observers in recalling results at some later time.

An Instrument of Discovery. One hundred years ago, Heinrich Hertz stated that:

> There is no escaping the feeling that these mathematical formulations (the models) have a life of their own, that they are more knowledgeable than those who discovered them and that we can extract more science from them than they contained originally.

A pictorial representation of results, generated from an underlying mathematical model, can exhibit visual forms whose structure and regularity may serve as a essential pointer to new areas for investigation (see Figure 7.12). Even in the most abstract fields (e.g., pure mathematics), virtual experimentation may lead to new discoveries (see Figure 7.6). Just as the microscope showed us the "infinitely small" and the telescope showed us the "infinitely large," so the computer will enable us to regard our world in a new, richer way.

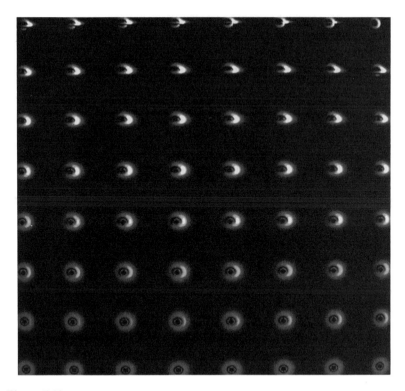

Figure 7.10

The theory of General Relativity proposed by Albert Einstein in 1915 is the framework for the description of the biggest scales of the universe. The theory makes provision for the existence of objects whose mass is such that nothing, including light, can escape from them (except in a quantum context). These objects cannot be directly observed, yet it is possible to describe characteristic signatures they would give rise to, such as a "gravitational lens" effect. For example, rays of light from a distant star will be deviated (as if in an optical lens) if they strike one of these objects on their path to Earth. The curvature of their trajectories causes a perturbation in their source image. The numerical experiment presented in this image shows 64 stages in the movement of a fictitious observer describing a circular trajectory taking him from the vertical of one of the poles of a black hole (bottom left) to his equatorial plane (top right). An accretion disk with circular symmetry gravitates around the equatorial plane. When the observer is at the pole, the disk appears as it would around a less massive object (e.g., Saturn), although secondary concentric images are already perceptible. During orbital movement, however, the most "disturbing" optical phenomena begin to occur. With this knowledge at his disposal, an astrophysicist can perform virtual experiments on these "cosmic abysses" in a way that could never be envisaged directly. (Model: Jean-Alain Marck, Observatoire de Meudon.)

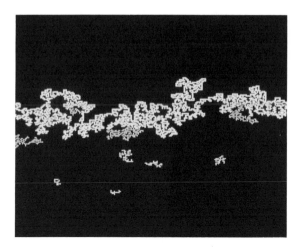

Figure 7.11
Diffusion representation.
During a simulation of
two-dimensional diffusion of
particles on a square mesh,
using random walk [6], the size
of the subsets of isolated
particles (including some that
are marked in pale yellow) is of
the same order as the thickness
of the boundary (marked in
white) in its vicinity. It then
decreases inversely to the
distance. (Model: Jean-Francois
Gouyet, Michel Rosso, and
Bernard Sapoval, PMC/X.) See
insert for color representation.

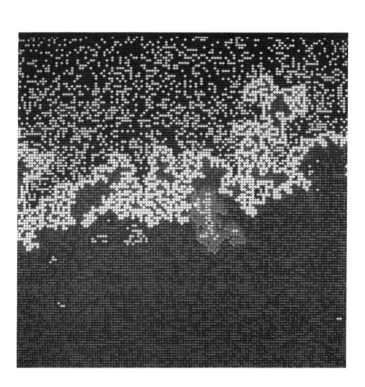

Figure 7.12
Diffusion representation.
During a simulation of
two-dimensional diffusion (see
Figure 7.11), a microscopic
event, namely the elementary
jump of the particle marked in
white, has macroscopic
consequences: The subset of
isolated particles marked in red
connects to, or disconnects from,
the boundary marked in yellow.
This being so, the measurement
of the length of the boundary is
subjected to variations of an
order much greater than that of
the size of the element
responsible for the
phenomenon. (Model:
Jean-Francois Gouyet, Michel
Rosso, and Bernard Sapoval,
PMC/X.) See insert for color
representation.

7.4 PICTURE SYNTHESIS: AN ARTISTIC TOOL

The computer is increasingly being used as a creative tool in the hands of artists. The computer and a high-resolution color display act as the brush and canvas of a traditional artist. One can readily imagine artists implementing within their computer programs the rules and techniques which have evolved through several centuries of painters in seeking to provide the desired emotional impact and creative style expected in paintings. As these rules and techniques are converted to computer-aided art tools, they may become established features of the computer programs used by scientists to display the results of virtual experiments. Perhaps it will be possible to remove the individual instincts and creativity of an experienced artist from the task of creating an appropriate scientific display of virtual experiments. Even if it is not possible to eliminate the artist as a member of the virtual experimentation team, the computer-aided tools developed by artists are likely to provide suitable tools for scientists to experiment with various ways of displaying numerical results.

For example, testing, comparing, retrieving, transforming, combining, memorizing, modifying, and deleting are just some of the artistic tasks that might be readily performed by a computer. Colors and forms become plastic and pliable, allowing the artist to change them, change his mind and return to an earlier point in his creative process, or simply delete everything. In addition, the sequence of steps followed by an artist starting with the bare "canvas" and progressing to the finished "painting" can be saved within the computer. Perhaps in some distant future, students will be able to study not only the final works of art of "classical" artists working today but also study the development of those famous paintings. These techniques, though of interest to artists, are also useful for the scientist seeking to develop his or her final visual display of the numerical results of a virtual experiment.

7.5 PICTURE SYNTHESIS: A NON-NEUTRAL TOOL

Creating computer graphics from abstract numerical results usually requires active, subjective decisions on the part of the scientist conducting the virtual experiment. The synthesis of pictures imposes several limits. Because video displays are limited to two-dimensional displays (at least presently), objects with dimension greater than two must be projected onto the display, perhaps providing an illusion of more dimensions (with graphics techniques such as perspective, hidden surface elimination, shading, anaglyph methods, and stereoscopy.). Motion may also appear in the visual presentation: A system may display dynamic behavior or may allow the observer to "walk around" the scene to study it from various viewpoints.

It is obvious that numerous artifacts are introduced when transforming numerical data to a visual display (e.g., aliasing, optical illusions etc.). Pictures resulting from virtual experimentation, like any other experimental result, must therefore be regarded with a degree of suspicion, and any surprising result must be initially be regarded as an error in the "experimental measurements."

In our context, *synthesis* is a process through which the separate elements (e.g., the individual numerical results of a virtual experiment) pertaining to a given object (e.g., the system being simulated during the virtual experiment) are gathered together into a coherent global representation (e.g., the image on the display). Synthesis of pictures will therefore be a set of techniques used to pass from the space (position and time) of the numerical results to the space (pixel and time) of display. The temporal and spatial concerns are discussed next.

The Temporal Factor: The image on the display can change with time, allowing picture dynamics to be used for data representation. Three rather distinct varieties of time can be used.

T_p corresponds to the inherent model time – for example, the variable t included in many equations.

T_m corresponds to the time while the viewer is "moving" around the objects being displayed. This is enormously helpful during the observation of complex systems with a spatial dimension greater than two.

T_d represents the case when the time variable is used to display a dimension of the results which cannot be displayed directly on a static screen. For example, the graph of a function $y = f(x)$ can be represented in the display in the traditional manner as a curve showing the value of the ordinate y for some range of values of the abscissa x within a Cartesian coordinate system. Alternatively, in a less conventional manner, the variable x can be represented by time in the display (i.e., $x \rightarrow t$, resulting, for example, in a mobile point moving with display time t on a single axis y of the display).

It quickly becomes obvious that these three different times can be used simultaneously, and it is clear (as we shall see later) that a number of precautions must be taken inasmuch as all three time factors refer to the same physical time.

Space of Results: The objects of a virtual experiment will generally be in some subspace $\mathbf{E}$ of a higher n-dimensional space $\mathbf{Q}^n$. The results generated by the virtual experiment will vary throughout the object and will generally be considered as functions $f(\mathbf{E})$ defined in each point of $\mathbf{E}$. Depending on the situation, the variables represented by the numerical results can represent different kinds of mathematical quantities, as seen in the following examples.

- The function f may represent a *scalar* variable, having a single numerical value at each point (e.g., a pressure field).

- The function f may represent a *multiscalar* variable, consisting of a set $\{f_i\}$ of functions, each function representing a different variable (e.g., a pressure field and a temperature field).

- The function f may be a *vector* quantity, representing the components of some vector variable at each point (e.g., a motion field).

Beyond these simple examples, it is possible to imagine more exotic functions (e.g., by overlaying a curl field with a motion field). In fact, it is clear that the above list is far from complete and is given solely to show the complexity of the problem. It is easy to convince oneself, as we shall illustrate below, that scalar functions in one- or two-dimensional space are the easiest to represent and are highly instructive, leading the way to a wide range of possibilities. In particular, a video display essentially presents a two-dimensional scalar field and the results of a virtual experiment must be mapped onto that field. It is then easy to see that visual synthesis (1) usually reduces the quantity (and quality) of the raw data results produced by a virtual experiment and (2) can present only one point of view of those results at a given moment in time (this remark does not ignore the notion of "windowing" but underlines the two-dimensional aspect of the display medium).

The display of results will typically require two distinct stages in mapping the raw results to the display screen's image. *The chain of pri-*

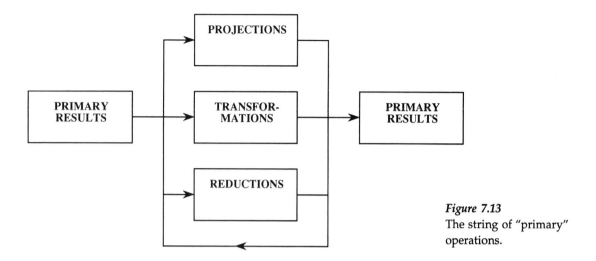

Figure 7.13
The string of "primary" operations.

mary operations provides the changeover from the space of unprocessed or *primary* results (i.e., results provided directly by simulation) to the space of two-dimensional *secondary* results (see Figure 7.13) which can be represented on a display screen. *The chain of secondary operations* is used to convert two-dimensional numerical quantities into the picture, and consists, in particular, of selection of the color palette.

7.6 PRIMARY OPERATIONS FOR DISPLAY OF NUMERIC RESULTS

Let us now look at the common operations used to pass from the space of the results of virtual experiments to the two-dimensional viewing space of the display [2].

Projection operations are used to pass from the n dimensions of the space of the results to the two dimensions of a plane (e.g., the space of the display). To view the entire space of the results of the virtual experiment, the viewer may need to "observe" the displayed object from various vantage points (i.e., perspectives), perhaps by "walking around" the displayed object to see opposite sides and hidden features. In this case, the inherent movement of the observer according to the time T_m, with times T_p and T_d frozen, can be very useful. Techniques used in flight simulators (a static, three-dimensional world through which the observer flies and observes) provide one familiar example of the general techniques which might be used. Suitable projections

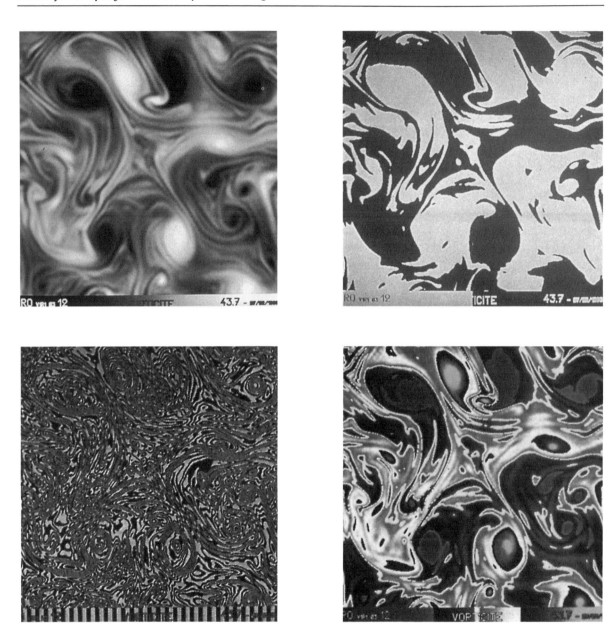

Figure 7.14
The simple case of a two-dimensional scalar field presented in four widely differing color palettes. (Model: Marie Farge, LMD/ENS.) See insert for color representation.

of three-dimensional fractal aggregates (see Figure 7.2) are often necessary to comprehend the features of the displayed "objects."

Transformations concern manipulation of results without changing the dimension of spaces. Such transformations might be used, for example, to reduce the dynamics of a field or, on the contrary, to enhance dynamics in certain important or interesting areas. It will be possible to differentiate between operations which transform the space of the results (e.g., translations, rotations, similarities, etc.) and those which transform the results themselves (e.g., the changes in dynamics mentioned above). In all cases, it is important that the picture resulting from such transformations contain data recalling the type of nontrivial operations performed on the original results to obtain the transformed results.

Reductions involve operations which typically simplify the data and include operations such as averaging, thresholding, and filtering. The latter is particularly useful in extracting the main structures from a field (see Figure 7.14). Finally, *cutting* is used to extract a subspace of results (e.g., a plane) and will be useful in presenting results with a spatial dimension greater than two.

7.7 SECONDARY OPERATIONS FOR DISPLAY OF NUMERIC RESULTS

The *chain of secondary operations* is used to produce the picture and consists of the selection of representation modes, in particular the color palette. This final operation is often completed in an adhoc fashion, yet it is a vital part of the process as a whole and it requires the definition of, and compliance with, a certain methodology.

RGB (**R**ed, **G**reen, and **B**lue) additive synthesis is used at the present time in television to reproduce "subjective" colors, or colored visual sensations; other color spaces could be defined as required. Without some apriori guidelines regarding the nature of the numerical results being presented on a display, there are no specific colors associated intrinsically with that display (What color is a pressure field or a quark?). However, the following two general remarks may be helpful when coloring data.

Remark 1: We tend to associate *cold* colors (green, blue, etc.) with low, negative values and to associate *warm* colors (red, yellow, etc.) with strong, positive, or even dangerous values.

Remark 2: We tend to associate a direct correspondence between a monotonically increasing luminance (levels of gray between black and white) and monotonically increasing numerical values represented by luminance.

These two observations may not provide an automatic definition of the colors required, but they provide guidelines for the selection process and help avoid a number of frequent errors. It is, for example, fairly common to see, in scientific publications, pictures using an arbitrary color coding scheme (possibly creating a conflict between the picture's aesthetic impact and its impact on conveying information regarding the displayed results). Consider the following association of colors with numerical values.

values 0 - 9	yellow	(warm, light color)
values 10 - 19	red	(warm, dark color)
values 20 - 29	blue	(very dark, cold color)
values 30 - 39	cyan	(very light, cold color)
etc.		

A picture using this rather arbitrary coloring scheme cannot be "read" without the aid of the table above specifying the association between color and numerical value. Moreover, even with the table, the picture is difficult to read because the code contradicts the two "cultural" observations indicated above (*Remark 1* and *Remark 2*). The mind switches between two possible interpretations (e.g., values 0 - 9 are low, yet they are coded using yellow which is naturally associated with high values).

This example also reminds us that there is no one natural order among the various colors but rather a number of orders. For example, the order of colors in the rainbow (i.e., red, orange, yellow, green, blue, indigo, violet) differs from the order of colors in television (i.e., blue, red, magenta, green, cyan and yellow). Furthermore, "white" and "black," which we often wrongly consider to be "colors," are instead extremes in the levels of luminance.

This being so, we shall distinguish between two types of color palette.

Monochromatic palettes (drawing on the subjective effect noted above in *Remark 2*) are useful for scalar fields. In this case, a single color is used with varying luminance to provide a visually "realistic" subjective understanding of the values of the numerical results.

Polychromatic palettes (in this case drawing on the subjective effect noted above in *Remark 1*) are useful to distinguish between various zones. In this case, the color scheme provides separation of regions. The visual information provided by the coloring scheme can be supplemented by adding variations (monotonic, periodic, or peaked) in luminance within a given color.

As a simple example of the above, a two-dimensional vorticity field produced by research into turbulence [3] is shown in Figure 7.14. In this case, the two-dimensional space of the results bears a strong similarity to the two-dimensional space of the display (defined on a 512×512 grid) and, therefore, to the picture. Each pixel in the picture directly corresponds to a spatial point (on a two-dimensional mesh) at which the vorticity field is computed. The numerical values of vorticities are scaled to fit within the range of numbers from 0 to 255 (i.e., represented by an 8-bit binary number). The scale of displayable colors is similarly represented by an 8-bit binary number. This provides a direct correspondence between each 8-bit value of the vorticity field and an ordered member of the 256 possible colors. Beneath each of the photos in Figure 7.14, each of which illustrate the same field, there is a color scale indicating the color associated with [0,255], if read from left to right. This type of indication must be systematically included with every picture to provide the viewer with information on how to view the results.

For photograph 1 of Figure 7.14, a monochromatic palette is used: the high negative values are represented in dark colors, while the high positive values are represented in light colors. For photograph 2 of Figure 7.14, an "all or nothing" two-colored palette is used (note that the warm color, yellow, has been linked to negative vorticities while the cold color, blue, is linked to positive vorticities, in order to create the uncomfortable sensation due to the inconsistency with *Remark 1* above). Although both photographs concern the same field, the visual impression is very different and only the main spatial structures stand out. In this case, low-frequency filtering is in operation. In photograph 3 of Figure 7.14, the previous palette is periodic, and the high spatial frequencies become obvious. Finally, in photograph 4 of Figure 7.14, a polychromatic palette is used to comply with the two basic observations *Remark 1* and *Remark 2*. In particular,

- Increasing luminance (varying in three peaks) follows the progression of the numeric values (*Remark 2*),

- Green (a cold color) is linked to negative vorticity (*Remark 1*),

- Red (a warm color) is linked to positive vorticity (*Remark 1*),

- Gray (a neutral "color") restricts the visual conflicts between red and green (as defined later), and, finally,

- A yellow line marks out variations in level (in this case, zero vorticity).

As predicted above, shapes stand out from the pictures spontaneously. The visual display of the results clearly shows features (e.g., slight waves appear at the top left of photographs 1 and 4 of Figure 7.14) which would be difficult (or impossible) to locate by use of image processing algorithms or other feature extraction algorithms. In a real sense, the use of such pictures relies on the natural ability of our visual senses to detect and isolate subtle effects in scenes being viewed.

Associating colors with numerical values is just one way of representing numerical data. As shown in Figure 7.15, we may wish to use visual clues to provide the appearance of a three-dimensional scene on the two-dimensional screen. In this case, the value of the vorticity is represented by a third spatial dimension. On the three corresponding photographs, by playing with lighting and the angle of observation, the vorticity of the field presented in Figure 7.14 is made to correspond with the "orthogonal height" of the field, which then defines a three-dimensional surface. The highest points have the highest positive vorticity, while the lowest points have the lowest negative vorticity. Whereas it may be difficult to compare the numerical values of the field on two distant pixels using only color and luminance, it may be easier to make that comparison by means of altitude above a plane, as illustrated by this example. Moreover, the surface defined in this way could be colored using the conventional mapping technique and a second field (see Figure 7.16), which makes it easier to obtain correspondences and determine correlations. "Spatial" defects that are imperceptible in the original field are more readily seen (see Figure 7.17).

This simple two-dimensional case illustrates that there is no single natural solution to the problem of display and furthermore that precautions must be taken to avoid visual responses which improperly interpret the displayed results (see *Remark 1* and *Remark 2* given earlier).

Finally, the *dynamics* of the numerical results describing the behavior of the system being studied using virtual experimentation must be considered. T_p may be an intrinsic part of the model. In addition, certain temporal (and in some cases spatial) correlations and spatial pairings between neighboring structures (e.g., positive and negative curls) may only become apparent during motion. Rather than displaying static pictures, the display now presents a "movie" of the

Figure 7.15

A three-dimensional representation of the two-dimensional field defined in Figure 7.14. See insert for color representation.

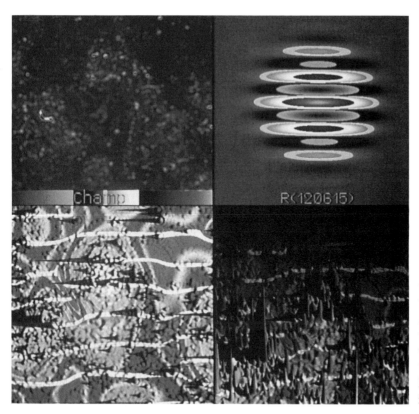

Figure 7.16

Wavelet transformation. A complex wavelet transform is a recent mathematical tool used for local analysis on all scales of a nonperiodic multidimensional signal (contrary to the Fourier transform). A field (top, left) is analyzed by the "Morlet" wavelet (top, right). The two views at the bottom display the field from different angles, as a three-dimensional surface onto which the transform module is 'mapped," while the zero phase lines are displayed in white. See insert for color representation.

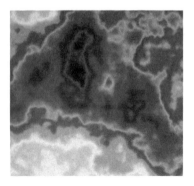 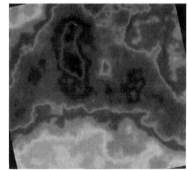 

Figure 7.17
Fractal field. A fractal field (1) is subjected to a rotation of $\pi/16$ using a "simplistic" algorithm. The two-dimensional presentation (2) shows no apparent defect while the three-dimensional display (3) indicates clearly that an anomaly exists (indicated by thick vertical lines and later explained by the absence of interpolation during the rotation process). See insert for color representation.

responses obtained from the experiment. Such effects are difficult to illustrate here – see Figure 7.18.

7.8 THE DECEPTIONS

The visualization of numerical results carries with it the potential of misrepresenting those results. We shall call these potential misrepresentations the *deceptions* of display.

Any single image is merely a single observation, yet in any real experimental situation several observations are typically needed to understand the behavior of the experimental system and confirm the validity of the experimental measurements. Therefore, several points of view are generally required (see Figure 7.14 and 7.15), perhaps using different representation modes and various color palettes. Unfortunately, those displays of results which will be useful are usually not known before experimentation begins, and, as a result, it may be necessary to collect results from a large number of experiments and on various ways of visually displaying the results. Advanced visualization systems are obviously important in completing these investigations efficiently.

Whether they appear within the primary or secondary operations performed on results, transformations are **not neutral**. In particular, they may weaken essential elements or, conversely, strengthen insignificant details (see Figure 7.19). Again, we are reminded that the virtual experimenter must be careful when developing visual displays of results to ensure that the real features of importance within the results are properly mapped onto features of importance on the display.

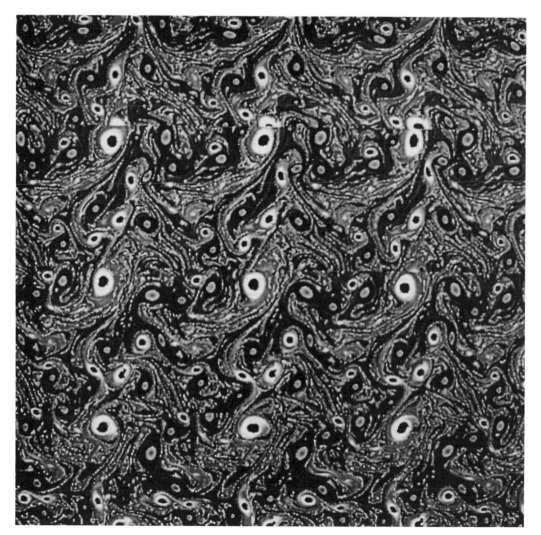

Figure 7.18

Turbulence. Certain spatial correlations can only be evidenced by an animated sequence of figures. These 16 images show 16 instants in a simulation of two-dimensional turbulence (the initial instant is at the bottom left, and the final instant is at the top right). The *red* (resp. *blue*) color characterizes the curls turning in the clockwise (resp. anti-clockwise) direction. Movement makes it possible to observe that, when two curls with opposite directions of rotation pass in the vicinity of each other, they form a structure known as a *modon*, which is stable in time and moves faster. (Model: Marie Farge, LMD/ENS.) See insert for color representation.

Figure 7.19
The Mandelbrot set connectivity. Are all parts of the Mandelbrot set connected? Judicious use of colors is important to illustrate the connectivity. If colors are poorly chosen, regions appear to separate from one another. A. Douady and J.H. Hubbard have recently proved that all parts are connected by thin filamentous strands (see Figure 7.6 for a definition of the Mandelbrot set). See insert for color representation.

A variety of *artifacts* are introduced into a displayed representation of results. The digital nature (both the discrete set of mesh points represented by screen pixels and the temporal sampling associated with successive frames of the display) of the displayed results introduces the common problem of aliasing (both spatial and temporal) due to excessively high frequencies in the results generated by the virtual experiment. The absence of absolute references for colors and luminance with regard to their association with numerical values has been discussed earlier. In addition, there are more subtle effects, in particular the *optical illusions* (Mach bands, simultaneous contrast, etc.). Such optical illusions are illustrated later. It is evident that a visual display, used to present the results of a virtual experiment, is not a neutral tool. Unfortunately, picture synthesis is often considered to be a neutral tool, and the use to which it is put often confuses the objectives of being aesthetically pleasing and being an informative representation of results. Let us remember that a picture produced in this way is, in this context at least, usually arbitrary and that, in fact, using the same set of data, a large number of very different representations can be constructed.

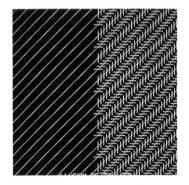

Figure 7.20
The Zöllner illusion (the long bars sloping at an angle of 45° are parallel, but they appear to undulate on the right side of the illustration). See insert for color representation.

The eye (or to be more precise the entire visual system) is susceptible to optical illusions, many of which are wellknown but too often forgotten. There are four categories of illusion, each of them responsible for a failure to appreciate correctly what is being seen.

1. *Geometric illusions.* These cause geometric deformities in the perceived shapes. In Figure 7.20, for example, the Zöllner illusion

introduces waves on lines that are objectively straight and parallel.

2. *Luminance illusions.* This type of illusion shows quite clearly that the eye has no absolute reference for luminance (luminance represents the intensity of gray at a given point, its minimum representing the intensity of black and the maximum the intensity of white). This being so, two zones with identical luminance separated by several other zones with a very different luminance will appear to be dissimilar (see Figure 7.21). It is, therefore, a total illusion to attempt to interpret numeric results quantitatively and generally if the results are represented by sets of luminances. Major relative errors might well be committed but, more seriously, inversions in the orders of size might be introduced. Moreover, when the luminance gradient changes sign, superluminous bands (the so-called Mach bands) appear (see Figure 7.22). Lack of knowledge of this phenomenon might result in erroneous conclusions about a numeric field.

3. *Chrominance illusions.* Luminance illusions have already shown that the visual system functions globally rather than locally. This being so, the neighborhood of a given pixel (or even the entire picture) affects the perception of the pixel itself. The same is true of chrominance. Figure 7.23 presents two fields with complementary coloring as regards the red and green. The zones colored gray, however, are identical in both displays as regards spatial localization and also, contrary to appearances, as regards actual color (i.e., the color allocated to them by a program). It is obvious on this figure that the "subjective" color of a zone is strongly influenced by the color of adjacent zones.

As to the black-and-white broadcasting of a color picture, it can occasionally give rise to rather unpleasant surprises. The color scale presented in Figure 7.24, for example, appears to have nonmonotonic luminance, whereas its black and white version ("calculated" using a television encoder) shows that this is not the case.

4. *"Computing" illusions.* The previous three categories of illusion were known to physiologists and those interested in art long before the arrival of computers. The fourth and final category, however, is much more recent and arises from the notion of false color. Most display systems in use at the present time provide

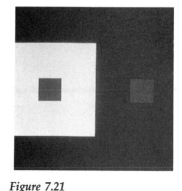

Figure 7.21
A simple contrast illusion. The two small gray squares have exactly the same luminance but appear to be different. Using a computer program, the observers vary the luminance of one of the squares by an average of 16% in order to bring it up to the level of the other visually. See insert for color representation.

Figure 7.22
Mach bands (the point at which the luminance gradient changes sign), where "superluminous" bands appear. See insert for color representation.

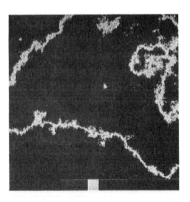 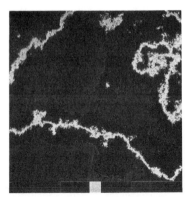

Figure 7.23
The same gray shade appears colored, depending on the neighboring color (although it is programmed to be identical in both images). See insert for color representation.

a means of coloring and re-coloring a picture. A digital picture in false colors can be defined as a matrix of pixels with each pixel representing integer coordinates X and Y. With each of these pixels there is an associated numeric value L (e.g., coded as an 8-bit data byte). When the displayed image is generated using this video image structure, the numerical strength code L of each pixel is converted to a color code C. This conversion involves the use of a look-up table which establishes the translation from the numerical strength code L to the color code C. For RGB monitors, the color code provides a triplet (C_r, C_g, C_b) representing the intensities of red, green, and blue, respectively. In this manner, the numerical value L of a pixel is "painted" on

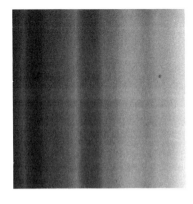 

Figure 7.24
The "real" luminance (2) of a color image can be very different from its "subjective" (1) luminance. See insert for color representation.

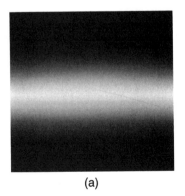

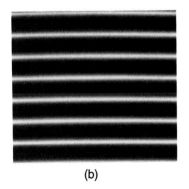

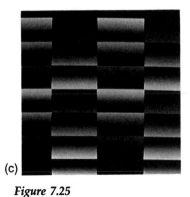

(a)   (b)   (c)

Figure 7.25

A "computing illusion" caused by using different color look-up tables. Depending on the colors used, incompatible conclusions are drawn from these three images, whereas the numeric field displayed is identical in all three cases. See insert for color representation.

the screen according to the (C_r, C_g, C_b) intensities. This means that identical results in the video memory (i.e., the matrix of L values) can be displayed differently depending on the contents of the color look-up table. This is one of the major dangers, but it is also one of the least recognized and most underestimated. Figure 7.25 shows three choices for the color look-up table, using the same data in the image memory for each picture. Each of three different color palettes can be used to draw a number of conclusions (all of them unfortunately incompatible). Figure 7.25a apparently shows two characteristics of the field. Horizontally, it is uniform; vertically, it increases and decreases. Note in passing that no quantitative data can be deduced from this observation. In particular, it is impossible to define whether the increase (resp. decrease) is linear, sinusoidal, or Gaussian. As for Figure 7.25b, it also indicates horizontal uniformity, but this time there are undulations in the vertical plane. Finally, in Figure 7.25c, there is a notable variation in the horizontal plane. The variations in the visual impact provided by different color schemes is clear.

From the above considerations, it is possible to draw a number of conclusions. First and foremost, there is no single, a priori mode of representation. Several modes (perhaps complementary modes) may be required to avoid interpretation errors or to comprehend various aspects of the "objects" represented. The second conclusion is that, contrary to a widely held belief, the display of results from numeric simulations is an "art form," drawing upon the subjective visual response of the viewer to the "painting." Coloring is not a casual, adhoc final step in a virtual experiment. Instead, it is a complex and integral part of the overall virtual experiment, providing the experimenter and his audience with a compact view of the behavior of the system being investigated. As we have seen, when carried out casually, improper coloring can stimulate conclusions by the viewer that are contrary to the actual numerical results obtained. Even if carried out with great care, it is important to remember that an aesthetically attractive picture may not be the best picture to provide an understanding of the numerical information being displayed.

7.9 THE FUTURE

As far as display itself is concerned, enormous progress will be seen, of course, thanks to the increasingly rapid manipulation of three-dimensional models and the development of high-definition television (HDTV) and stereoscopic devices. The notion of *virtual reality*, which has been developed in other fields (driving simulators for example) and introduced into this context, will enable researchers, engineers, or even artists to become more closely integrated into their models, and this in turn will facilitate comprehension. It should be remembered at this juncture that, unlike applications of the CAD (computer-aided design) type, the objects that are studied and displayed are often far removed from current understanding; indeed, many of them have no natural image at all (pressure, for example). Representation of some of them may even be prohibited (in particular in quantum mechanics), and in this case a scientist must be given the maximum facilities in order to understand the results obtained. In this field, the techniques of artificial intelligence seem likely to be of use.

It is difficult to project the impact of data processing and scientific visualization on the field of scientific research. Science defines our rational view of the universe and the place that man occupies within that universe. These views have undergone major changes over the centuries as the result of new insights and techniques developed by Plato, Copernicus, Newton, Einstein, and many others. At present, a new generation of scientist is developing tools and drawing on computers and visualization to continue the development of scientific research. These new tools may lead to a new "Copernican revolution" in the future.

REFERENCES

[1] A. Bouhot, Le Pouvoir Créateur des Mathématiques, *La Recherche,* 1340-1348, (1990).

[2] J.D. Foley and A. Van Dam, *Fundamentals of Interactive Computer Graphics,* Addison-Wesley, Reading, MA (1984).

[3] M. Farge and R. Sadourny, Wave-Vortex Dynamics in Rotating Shallow Water, *J. Fluid Mech.* **206** 433-465 (1989).

[4] B. Mandelbrot, *The Fractal Geometry of Nature,* Freeman, New York (1983).

[5] J. F. Colonna, Animation of fractal objects, in *Computer Graphic Interface '89,* Canada, (June 1989).

[6] B. Sapoval, M. Rosso, J.F. Gouyet and J.F. Colonna, Dynamics of the Creation of Fractal Objects and 1/f Noise, *Solid State Ionics,* North-Holland, Amsterdam (1986).

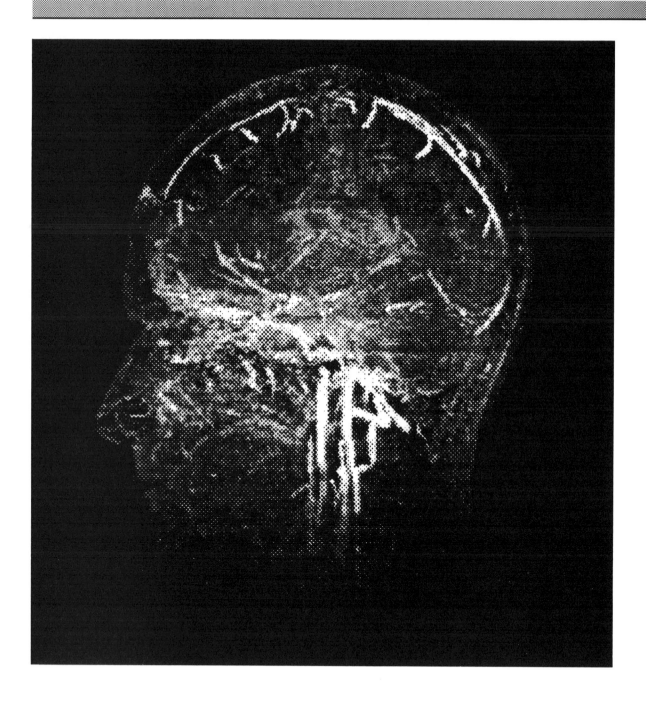

Architecture and Applications of the Pixel Machine

Michael Potmesil

AT&T Bell Laboratories
Holmdel, New Jersey

and

Eric M. Hoffert
Apple Computer, Inc.
Cupertino, California

8.1   INTRODUCTION

As computing technology progresses, it becomes apparent that even the most powerful computers available, built on principles devised by John von Neumann in the early 1940s, are reaching the limits of their speed imposed by the constraints of physical laws. The single-processor model executing only at most one instruction in every machine cycle is beginning to outlive its usefulness. This model forces many problems of different structures to fit into one particular paradigm. This paradigm is not necessarily natural for algorithm development or computational efficiency: many algorithms have large parallel components to them. There is no inherent reason why many calculations cannot be performed simultaneously. Computer graphics is a perfect example of such an application area. Pixels can be read, written and processed simultaneously; in fact, most graphics algorithms impose few limits on the amount of parallelism achievable for pixel processing. Based on these observations, the Pixel Machine was designed and built as a specialized computer graphics machine with a *parallel* architecture.

With a parallel architecture, a designer hopes that, instead of the typical linear improvement in performance that is inherent in technology evolution, a quantum leap in performance can be obtained. Such a quantum leap has been *demanded* by the various communities using image computing. The recent report *Visualization in Scientific Computing* [1] stresses the need for innovative high-speed architectures to meet the needs of interpreting large amounts of scientific data. Animators require photorealistic rendering of high scene complexity and image quality with quick turnaround times. Doctors and radiologists must see a three-dimensional reconstruction from an nuclear magnetic resonance (NMR) or computerized tomography (CT) device in seconds. For image computing to be a practical tool in these and other areas, it is not feasible to wait for evolutionary improvements in technology. Instead, a break from traditional architectures must occur and be built. In this chapter, we describe such an architecture; what motivated its development, how it works, and what it portends for the future of image computing.

The Pixel Machine[1] is a parallel image computer with a distributed frame buffer. Its architecture is based on an array of asynchronous MIMD nodes with parallel access to a large frame buffer. MIMD stands for "Multiple Instructions Multiple Data." It means the machine consists of a number of processors, each of which runs its own instructions (MI), and also each of which works on separate data (MD). Recently, the term MIMD refers primarily to parallel machines composed of many 'nodes," each consisting of a microprocessor, memory, and means to communicate with other nodes by sending messages. Examples include the Intel Paragon, Ncube systems, IBM's SP1, and Thinking Machines' CM-2 (but not CM-1). Each node has its own code and executes it. The primary alternative is "SIMD," in which each node is not a whole microprocessor able to execute its own in-

---

[1] The Pixel Machine was one of the early graphics accelerators using parallel processing (with high speed digital signal processors). Experimentation with the Pixel Machine helped define many of the architectural principles which continue to remain important.

structions; rather, instructions are broadcast from a controller to all nodes, which all do the same thing.

The machine consists of a pipeline of *pipe nodes* which execute sequential algorithms and an array of $m \times n$ pixel nodes which execute parallel algorithms. A *pixel node* directly accesses every mth pixel on every nth scan line of an interleaved frame buffer. Each processing node is based on a high-speed, floating-point programmable processor. We present the mappings of a number of geometry and image computing algorithms onto the machine and analyze their performance. Examples of the machine being used as a general-purpose parallel computer are also given.

The design of the system architecture was influenced, inspired, and motivated by the following concepts.

Speed: The advent of fast RISC-style digital signal processors that offer a large amount of the functionality found in a microprocessor with an integrated floating-point unit at a fraction of the price [2].

Parallelism: The hypercube architecture [3], in which processing is performed in parallel by nodes on the contents of their own local memories and messages can be exchanged between processors.

Interleaving: The notion of an interleaved frame buffer to achieve load balancing in a parallel image-computing system as originally developed in [4,5].

Programmability: The concept of a programmable graphics machine attached to a host computer as introduced in the *Ikonas* frame buffer and graphics processor.

Pipelining: Pipelined operations as applied in the Geometry Engine[2] to geometry computing [6].

Flexibility: The value of a rendering and modeling programming environment such as FRAMES [7], where different computing modules following the old software adage *"small is beautiful"* can be interconnected in different ways to achieve diverse modeling and rendering functions.

Partitioning: Image-space or object-space partitioning of data among 2D or 3D arrays of asynchronous, independent processing elements as described in reference [8].

---

[2] Geometry Engine is a trademark of Silicon Graphics, Inc.

The design of the machine was, of course, simultaneously constrained by a number of practical and very much intertwining factors such as cost, memory capacity, physical design, component availability, size of cabinet, number of cards, size of cards, size of components, and heat dissipation. We also had to consider performance tradeoffs between a larger number of simple nodes and a smaller number of more complex nodes.

8.2   SYSTEM ARCHITECTURE

The Pixel Machine was designed as a programmable computer with pipeline and parallel processing closely coupled to a display system. The computer consists of four major building blocks (see Figure 8.1):

1. A pipeline of *pipe nodes,*

2. An array of $m \times n$ parallel *pixel nodes* with a distributed frame buffer,

3. A *pixel funnel*, and

4. A *video processor.*

Figure 8.1
A block diagram of the Pixel Machine.

The pipeline and pixel-array modules can be incrementally added to a system to build a more powerful computer. The Pixel Machine func-

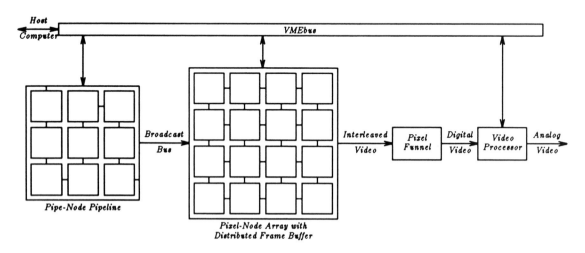

tions as an attached processor. In the current configuration the host computer is a high-end workstation, but, in principle, diverse hosts could be supported, ranging from personal computers to supercomputers.

8.2.1 Computations

The CPU of the pipe (Figure 8.2) and pixel (Figure 8.3) nodes is a DSP32 digital-signal processor with an integrated floating-point unit [2]. It consists of a 16-bit integer section and a 32-bit floating point section. The integer section with 21 registers is mainly used to generate

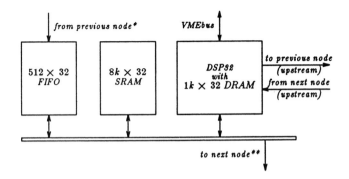

Figure 8.2
A block diagram of a pipe node. * From the VMEbus in the first node of a pipeline. ** To the broadcast bus and the VMEbus in the last node of a pipeline. VMEbus is a registered trademark of the VME Manufacturers Group.

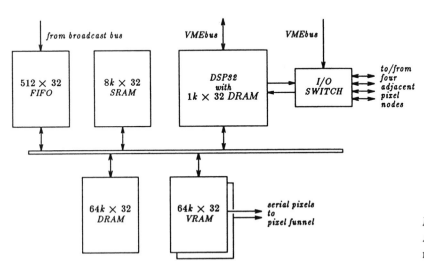

Figure 8.3
A block diagram of a pixel node.

memory addresses, while the floating-point section with four 40-bit accumulators is used to process geometry and image data.

The DSP32 has a RISC-style instruction set and instruction decoding. Unlike a RISC processor, which operates only on data in registers and uses load/store register-memory accesses, the DSP32 uses register pointers to point to arrays of data in memory. The pointers are usually post-incremented during the same instruction. In a typical operation, the DSP32 can read two operands from memory and one from an accumulator, perform a multiply-accumulate operation, and write the result to an accumulator and to memory.

The DSP32 has a 16-bit addressing capability, allowing it to address directly only 64 Kbytes of memory.[3] There are 4 Kbytes of RAM memory on board of the chip. Each pixel and pipe node has additional 32 Kbytes of fast static RAM memory. These 36 Kbytes are used for program and scratch data storage.

Pixel nodes also contain a distributed frame buffer and z-buffer. In each pixel node, there are 512 Kbytes of video RAM memory organized as two banks of 256×256 32-bit $rgba$ pixels and 256 Kbytes of general-purpose dynamic RAM memory which can be organized as a 256×256 32-bit floating-point z-buffer. These additional 3/4 Mbytes of memory are addressed via a memory management unit.

The nodes are running at 5 Mips or 10 Mflops, which must really be interpreted as 5 million multiply-accumulate operations per second. In typical applications, programmed in C, the overhead of invoking functions, computing data pointers, and so on, can reduce the floating-point operations to about 10-25% of the peak rate.

8.2.2 Communications and Connections

There are a number of different communication paths in the system. Each pixel and pipe node is connected to the VMEbus via a DMA port (*host-to-node connection*). This port can be used by the host to access all memory-mapped locations in a node and for handshaking and synchronization activities by the node.

Pipe nodes are connected with fifos into nine-node pipelines (*downstream pipe node-to-node connection*). The fifo input to the first node is written by the host via the VMEbus, and the fifo output of the last node is either broadcast (via a broadcast bus) to all the pixel nodes (*pipe-node to pixel-node connection*) or written to a fifo read back (via the

---

[3]It should be noted that the next generation of this processor has a 24-bit addressing space allowing it to address directly 16 Mbytes of memory.

VMEbus) by the host. The pipe nodes in a pipeline are also connected via a unidirectional serial asynchronous link in direction opposite to the fifos (*upstream pipe node-to-node connection*). Two pipelines can be placed in a system and configured as two parallel pipes or one long serial pipe.

Pixel nodes are connected to their four nearest neighbors, in a closed torus network, via serial bidirectional asynchronous links (*pixel node-to-node connection*). These pathways allow flexibility for data movement needed in different algorithms.

8.2.3 Pixel Mapping and Display

The frame buffer in the Pixel Machine is distributed into the array of the $m \times n$ pixel nodes. The frame buffer is divided into two or more buffers. One of these buffers is always displayed by the video system, at the selected refresh rate, on the screen. When in double-buffered mode, a second buffer is used to draw the next image. Additional buffers may hold texture maps, volume data, or invisible parts of windows. Pixels in the frame buffer being displayed are read by a video processor and mapped on the video screen. This mapping is determined by the position of each pixel node within the array and is fixed. Each pixel node contains the size of the pixel-node array (m, n) and its position within the array (p, q), where $0 \leq p < m$ and $0 \leq q < n$. The position (p, q) also serves as a unique identification number of each node. Pixel node (p, q) then displays every mth pixel starting with pixel p on every n-th scanline starting with scanline q; that is, a processor-space pixel (i, j) is mapped into a screen-space pixel (x, y) by

$$x = im + p \tag{8.1}$$

$$y = jn + q \tag{8.2}$$

This format requires the display subsystem to collect all of the distributed frame-buffer pixels and assemble them into a contiguous screen image. This function is performed by the video *pixel funnel*. The interleaved format of the frame buffer provides load balancing in image-computing algorithms, and it matches well the speed limitations of video RAM memories with the speed requirements of a high-resolution display.[4]

---

[4] A pixel is shifted out of a video memory in ≈ 40 ns, whereas it is displayed on a 1280 x 1024 pixel screen in ≈ 9 ns. Therefore, at least 5 parallel banks of video memories are required to shift out 5 pixels in ≈ 40 ns.

Table 8.1: Physical and Virtual Pixel-Node Configurations

Physical			Virtual			V/P
Nodes	m × n	Pixels/node	Nodes	m′ × n′	Pixels/node	Ratio
16	4 × 4	256 × 256 *a*	64	8 × 8	128 × 128	4
20	5 × 4	256 × 256 *b*	80	10 × 8	128 × 128	4
32	8 × 4	128 × 256 *a*	64	8 × 8	128 × 128	2
40	10 × 4	128 × 256 *b*	80	10 × 8	128 × 128	2
64	8 × 8	128 × 128 *a*	64	8 × 8	128 × 128	1
		160 × 128 *a*			160 × 128	1

a Screen size 1024 x 1024 pixels.
b Screen size 1280 x 1024 pixels.

The architecture of the pixel nodes is scalable, using between 16 and 64 nodes (Table 8.1). To aid in the development of uniform software for all the pixel-node configurations and to allow hardware modularity, the concept of *virtual pixel nodes* was utilized. A virtual node renders into a subset of a buffer, called a *virtual sub-buffer*, all within a physical node. The virtual nodes and their sub-buffers are also interleaved in an $m' \times n'$ pattern - just as the physical nodes - with each virtual node having a unique screen position (p', q'). The mappings in equation (1) also apply to the virtual nodes. All software is written for one virtual node and is invoked one or more times, depending on the system size, by a physical node. The physical and virtual pixel-node configurations of the Pixel Machine are shown in Table 8.1.

8.3 SOFTWARE ENVIRONMENT

Software developed to run on the Pixel Machine is always divided into two major conceptual areas: *host software* and *node software*. The latter category is further subdivided into *pipe-node software* and *pixel-node software*. Host software controls interaction with the Pixel Machine, pipe-node software executes sequential-type algorithms, and finally pixel-node software executes parallel algorithms.

8.3.1 Host Software

Each node in the Pixel Machine is a small autonomous computer, albeit with a number of limitations. The current processor used in each node does not support interrupts and has limited addressing capabilities. These limitations forced the software designers of the Pixel Machine to come up with a number of creative solutions to difficult problems typically not encountered on a conventional sequential computer. A programming environment had to be developed that simulates much of the functionality taken for granted in a standard operating system. The following sections touch on some of the problems encountered along with their solutions.

There are two different types of processes which can run on the host computer and interact with or control the Pixel Machine:

Passive server: This process functions as a database server for the Pixel Machine. In this capacity, interaction takes place in a linear fashion: The host sends a stream of commands and data to the Pixel Machine, and the Pixel Machine performs various operations on the received data. There is no interaction initiated by the Pixel Machine, and it responds only when it is explicitly requested to do so (e.g., to a command to return the current transformation matrix). This server is employed almost exclusively for traditional polygon rendering, where databases and commands for each image are generated by the host and sent to the machine. In this mode the Pixel Machine acts as slave and the host computer as master.

Active server: This process is responsible for responding to all requests for resources that are made by the Pixel Machine. It polls a user-defined set of nodes (pipe *or* pixel) for messages. When a message is received, the active server initiates a host function that supplies needed resources to the requesting node. We have found this to be a very powerful paradigm for host/Pixel Machine interaction, and it is used quite frequently. Its power comes from the concept of sharing resources between a host computer and an attached processor. The host needs the Pixel Machine for certain demanding geometry and image computing, and the Pixel Machine needs the host for contiguous large blocks of memory and for access to a file system (among a number of other potentially possible needs). In this mode the Pixel Machine acts as master and the host computer acts as slave.

The host process has complete control over all nodes. It can access all memory in each node, including program memory and framebuffer memory in pixel nodes. Such accesses take place, via DMA, even when the nodes are running. The host software is also responsible for halting, initializing, and starting each node as well as for downloading programs into them. It also configures the video processor and accesses the video lookup tables.

8.3.2   Pixel-Machine Software

Software that runs on the Pixel Machine is quite distinct from software that runs on von Neumann machines. The important distinction from the single-processor approach is that software is mapped to different architectural components, each of which has a different *character* and number of nodes. The pipeline (where each pipe node typically contains a distinct program) executes sequential algorithms, and the pixel-node array (where each pixel node typically contains the same program[5]) executes parallel algorithms. In some cases, our algorithm is entirely sequential; such an algorithm would run only in the pipe nodes. Analogously, we have algorithms that are entirely parallel in nature; such an application might not utilize the pipeline at all. We have found that most applications have components that map onto both the pipeline and pixel-node array.

8.3.3   Pipe-Node Software

Pipe-node software requires algorithm partitioning. Each pipe node acts as a distinct computational element in a pipeline. A separate program runs in each node, and messages (commands and data) are passed down a pipeline. The last node in a pipeline has the ability to broadcast messages to all the nodes in a pixel-node array or to return them back to the host.

Pipe-node software can be written to allow the same software module to reside in several consecutive nodes and to operate on alternating input messages (e.g., each instance of an n-node transformation module transforms only every nth polygon). This allows the same software to run efficiently in longer pipelines and to eliminate or reduce bottlenecks by repeating the slowest module of a pipeline in more than one node. The end effect is that the pipe behaves like a vector processor.

Computation in Pipe Nodes

Pipe nodes compute on geometry, image, or volume data. They are employed for operations that are intrinsically sequential in nature. Such operations are the ones that constrain the efficiency of a parallel algorithm. The use of a pipeline is an attempt to remove as much

---

[5]However, there is no reason why each pixel node cannot execute a different program.

sequential style processing from the parallel pixel-node array as possible. Partitioning sequential algorithms among multiple nodes in a pipeline fashion decreases execution time further, above and beyond the decrease in execution time contributed by the parallel pixel-node array. The FRAMES system [7] contains methods for experimenting with this partitioning and on how to achieve maximum flexibility in such a scheme. The same philosophy is employed here. Our experience shows that special care must be taken to ensure that software in the pipeline does not become I/O bound. We would like to keep the computation/communication ratio as high as possible.

Communication Between Pipe Nodes

Each node in the pipeline can pass data to a subsequent node in the pipeline. Communication between nodes occurs via fifo devices which buffer messages as they are transferred from one node to the next.

8.3.4 Pixel-Node Software

This section describes (a) what actions a pixel node performs as a computational element and (b) the general mechanisms available for increasing the amount of data that a pixel node has access to, above and beyond the amount of memory fixed by hardware. There are two approaches to the issue of memory limitation. The first approach is that of message-passing, where nodes exchange portions of distributed data using some type of network (e.g., 2D grid, 3D grid, network, hypercube). This approach exploits the ability of a machine to shuffle large amounts of data among nodes very quickly. The second approach utilizes the memory of the host computer, letting it serve as an adjunct memory device for individual nodes. We discuss these two approaches in subsequent sections on message passing and virtual memory for pixel nodes.

Computation in Pixel Nodes

Computation in the pixel nodes comprises three categories: (a) frame-buffer and z-buffer access, (b) screen-space to processor-space coordinate conversion, and (c) optimized math functions.

Each node can independently read or write the contents of its individual frame buffer and z-buffer. Access to these memories is in row or column addressing modes. A 32-bit-wide pixel can be accessed in four instruction cycles (one cycle to read each color component from the frame buffer) and a 32-bit z-buffer value in one cycle.

Mapping functions transfer coordinates from the (x, y) screen space to the (i, j) processor space of a pixel node (p, q) by

$$i \;=\; \frac{x - p}{m} \tag{8.3}$$

$$j \;=\; \frac{y - q}{n} \tag{8.4}$$

Other functions perform the inverse mapping from processor space to screen space, map screen space linear functions such as $ax + by + c$ to processor space, and map screen space spans such as $(xmin, xmax)$ or $(ymin, ymax)$ to processor space.

Math functions include routines for frequently used operations in geometry and image computing such as square root (ray-sphere intersection), vector normalization (shading), and dot product (backface removal). These highly optimized functions efficiently utilize the floating-point capability of the DSP32 at each node, since much of the operations involve multiply/accumulate instructions.

Communication Between Pixel Nodes

Each node in the pixel-node array has a four-way serial I/O switch. This allows a node to communicate directly with its four nearest neighbors. Communications between two nodes occur over a half-duplex serial channel. All nodes must synchronize to exchange data, and message-passing occurs in lockstep fashion, with all nodes sending data in the same direction at the same time. This type of communication scheme is well-suited to problems that map onto a closed torus architecture such as block-iterative processes and image convolutions.

Interleave/De-Interleave

There are times when it is undesirable to compute on pixels in the interleaved format. Algorithms may require access to contiguous image tiles or complete raster lines or columns. Using the current Pixel Machine, this is not possible through hardware due to constraints imposed by video memory access requirements. At this point, the old hardware adage *"do it in software!"* is employed.

In this case, software is used that can take the interleaved frame buffer pixel data and, using serial I/O message-passing, reconfigure the frame buffer so that each node has a contiguous block of pixels. We call this process *de-interleaving*. Analogously, it is possible to take a frame buffer configured as contiguous blocks and, again employing serial I/O message-passing, distribute the pixels so that they are in their correct interleaved position for display. We call this method *interleaving*.

8.3.5 Virtual Memory

Parallel machines of the message-passing type typically have very few, if any, mechanisms for virtual memory access. This presents a problem for algorithms with large memory requirements. We implemented a number of different novel approaches for handling virtual memory. Because a general scheme for virtual memory was not possible with the current processor, several specialized methods were devised with reasonably good results.

Photorealistic rendering often requires large amounts of data. These data are typically geometry information, but can also consist of texture maps, environment maps, and so on. Other rendering techniques, such as volume rendering, can also require significant amounts of data storage. We have also found that an efficiently coded implementation of a rendering program (ray or volume tracers, for example) can be very small, in terms of code space. Hence it became apparent that we could develop schemes for virtual memory which would be used only for *data*.

Parallel Paging

On a distributed-memory machine, the same type of scheme used on a von Neumann machine for virtual memory [9] can be employed with a few twists. Each node has a page table in its memory along with a set of associated pages. When a memory access is required for data that does not reside in the available pages, a *parallel page-fault* is generated, causing a node to make a request to the host to deliver the required page of memory. The page is broadcast to all nodes in the pixel array from the last node in the pipeline. At this point, the page table in each node is updated, deleting a page based on the page replacement policy and adding the newly requested page to the table. We call it *parallel paging*, since typically nodes may request pages from the host concurrently.

Virtual Shared Memory

The active-server process can store multiple large texture maps or volume databases in main memory. When an individual pixel or voxel is requested by an arbitrary node in the system, the host retrieves the data requested from an array in main memory and routes it to the requesting node in the system. Our experience has proven that requests from nodes for pixel or voxel data should be collected into packets for optimal performance. This scheme is especially suitable for either (a) applications with memory requirements that far exceed the collective memory capacity of the pixel nodes, or (b) applications where distribution of memory over the pixel nodes would require an overly complex and/or inefficient algorithm. The method will be efficient as long as the host computer has a large amount of main memory. Because all pixel nodes have access to this memory, we call it *virtual shared memory.*

Program Overlays

As mentioned earlier, each node's processor can address 64 Kbytes of memory. This constraint, coupled with the cost and size of fast static RAM memories, dictated the size of program memory at 36 Kbytes in the current Pixel Machine. The solution to this problem of small program size is a classic one first seen in the early days of computing.

If a node does not have enough program or local data memory available for a required function or message processing, we use *program overlays* [9]. A program is manually divided into (a) a static instruction and data segment which resides in a node at all times and (b) several dynamic segments which are swapped-in one at a time from the host. The host server keeps track of the overlay segments loaded into any of the nodes and ensures that the correct segments are loaded into each node before data requiring them arrive. The cost of sending overlays from the host and loading them into a node's program memory is small: The bandwidth from the host to the pipeline is on the order of Mbytes/sec, and the overlay segments are on the order of single Kbytes. We can, therefore, change overlays several times within a frame while computing images at "real time" rates.

8.4 THEORETICAL PERFORMANCE ANALYSIS

In this section we attempt to analyze the theoretical performance of the Pixel Machine architecture and then look at some of our actual results while describing various image-computing algorithms.

The classic recurrence equation for the *divide-conquer-marry* paradigm is as follows:

$$T(n) = g(n) + MT(n/M) + h(n)$$

where $g(n)$ is the cost of dividing up a problem into M subproblems (*divide*), $T(n/M)$ is the cost of running the subproblem (*conquer*), $h(n)$ is the cost of combining the results of the subproblems into a final solution (*marry*), and n is the number of data elements. This generic equation is typically applied to a sequential implementation of a recursive algorithm. Interestingly enough, the equation can also be applied to the analysis of algorithms on parallel machines. In this case, the multiplicative term M would drop out, since the *divided problems* or subproblems are being solved concurrently. The modified equation becomes

$$T(n) = g(n) + T(n/M) + h(n)$$

The ideal parallel algorithm will have minimal $g(n)$ and $h(n)$ terms; these are the *parallel overhead* costs. The algorithm development efforts for parallel architecture are primarily concerned with ensuring that the $T(n/M)$ term will predominate in the expression above. This ensures that adding more processors to a problem yields a linear improvement in performance. A term that has recently entered into the parlance of parallel processing is *non-von Neumann bottleneck*. This refers to the costs $g(n)$ and $h(n)$, which are considered bottlenecks if they predominate in the expression above.

The salient difference between the Pixel Machine and other parallel machines is that there is no $h(n)$ term for displaying or animating the image computed by the pixel nodes. This immediately obviates a large amount of the usual parallel overhead. This term is eliminated because the interleaved frame buffer is assembled into a contiguous scan image by the pixel funnel within a video-frame time. The video hardware is essentially performing the *marry* function for "free." Thus, at least a large part of the non-von Neumann bottleneck is removed in this special case. Only if we read back the computed image from the frame buffer into main memory or secondary storage does the $h(n)$ term reappear.

The $g(n)$ term represents the cost associated with the screen space to processor space conversion. As an example of how this term affects efficiency, consider the case of rasterizing a geometric primitive in a pixel node. A simple equation describing the rasterization is as follows:

$$T(p) = g(x) + I(x)p$$

where p is the number of pixels rasterized, $T(p)$ is the time required to rasterize these pixels, $I(x)$ is the cost per pixel of rasterization for an arbitrary algorithm x, and $g(x)$ is the parallel overhead for that algorithm. Let us also define η, the efficiency of a parallel algorithm implementation, to be the slope of the graph of normalized inverted execution time versus number of pixel nodes. A unity value of η implies exactly linear improvement in performance for linear increases in the number of pixel nodes. This is what we aspire to for all implementations. Values less than unity indicate sublinear improvement for pixel-node increases. If p is small and $g(x)$ is large so that $g(x) > I(x)p$, then the parallel overhead predominates and $\eta \ll 1$. Conversely, if p is large and $g(x)$ is small so that $g(x) < I(x)p$, then the parallel overhead is small or negligible and $\eta \approx 1$. Measurements prove our analysis to be true. As an example, the Pixel Machine is extremely efficient when rasterizing very large polygons, and efficiency is lower when polygons are on the order of a few pixels.

The optimal algorithms for the Pixel Machine are those that require a $g(n)$ term only once per *image* as opposed to once per *object*. An example of the former is ray tracing, and an example of the latter is vector drawing. It is much easier to amortize the cost $g(n)$ once per image than once per object, since there usually are many objects in an image.

8.5   PARALLEL IMAGE COMPUTING

The flexibility of the Pixel Machine has been a powerful tool for algorithm development. In this section, we describe the mappings of a number of geometry and image computing algorithms to the Pixel Machine architecture.

8.5.1   Polygonal Rendering

Points, lines, polygons, and other geometric primitives are transformed, clipped, shaded, projected, and broadcast by the pipeline nodes. Complex geometric primitives (patches, superquadrics) are also generated or converted into polygons in the pipeline. The pixel-node array is used for raster operations, rendering of geometric primitives, z-buffering, texture mapping, image compositing, and antialiasing. For polygonal rendering, the passive server is employed, routing large polygon databases or multiple frames of animation to the Pixel Machine via the pipeline. Image antialiasing is accomplished by supersampling and floating-point convolution with an arbitrary filter kernel.

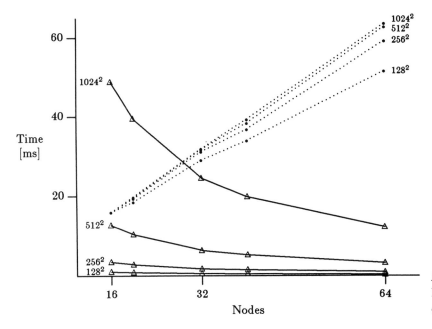

Figure 8.4
Parallel performance of raster
operations.

8.5.2   Raster Operations

A fundamental pixel-node function is to modify rectangular regions on
the screen in various ways. The pixel node organization allows $m' \times n'$
pixels to be processed by $m' \times n'$ virtual nodes during each iteration in
parallel. Figure 8.4 illustrates performance of the pixel nodes for raster
operations on 128^2, 256^2, 512^2, and 1024^2 pixel regions. The execution
times are plotted as solid lines, and the normalized efficiency is shown
as dotted lines. The efficiency of the machine, as the slopes of the
dotted lines indicate, is very high with almost linear improvement
and increases as the size of the region increases.

8.5.3   Line Rasterizing

A parallel version of the Bresenham algorithm rasterizes one-pixel-
wide lines. In an $m' \times n'$ array of virtual pixel nodes, the algorithm
writes $\min(m', n')$ pixels during one iteration. The figure of merit for
this algorithm is only $\min(m', n')/m' \times n'$. Line drawing, which is
essentially a one-dimensional process, cannot be very efficiently implemented on this architecture. Performance for randomly oriented
2-, 16-, 128-, and 1024-pixel-long lines is shown in Figure 8.5. Actual
times are plotted in solid lines, while the efficiency of the algorithm is

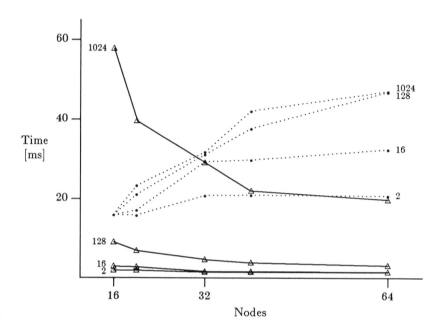

Figure 8.5
Parallel performance of line
rasterizing.

plotted as dotted lines. As expected, the slope of these lines illustrates
the low efficiency. For very short vectors the overhead becomes dom-
inant, and there is almost no improvement in speed as the number of
processors increases.

8.5.4 Polygon Rasterizing

A pipe node converts polygons into triangles, sorts their vertices in y,
and computes the slopes of the three edges. The pixel nodes transform
the slopes into their processor spaces, compute forward difference in y
along the edges, and scan-convert the triangle by stepping in y along
two active edges and filling the span in x between them. Performance
for random triangles within 8^2, 16^2, 64^2, 256^2, and 1024^2 bounding
squares is given in Figure 8.6. For large triangles the performance is
similar to raster operations. As the size of the triangles decreases, the
sequential part of the rasterizing algorithm dominates and decreases
the efficiency of the architecture.

Illustration: The airplane model (Figure 8.7) consists of 2116 bicubic
patches and 321 spheres. Each patch was tessellated by a pipe node
into 18 polygons. The vertices of the polygons were shaded using
two directional light sources by a shading pipe node. The polygons

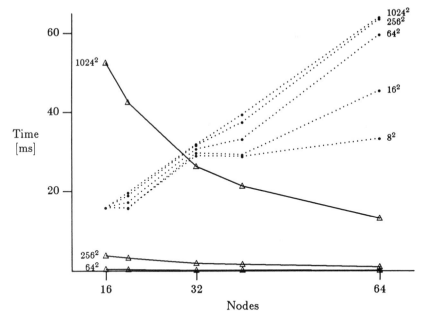

Figure 8.6
Parallel performance of polygon rasterizing.

Figure 8.7
Polygonal rendering: An alias airplane with procedural clouds over Dog Mountain, Colorado. See insert for color representation.

were then rasterized and filled with interpolated intensities by the pixel nodes. The rivets on the plane body were rendered directly by the pixel nodes using a sphere rendering function. The clouds were generated by a 2D solid texture technique. The terrain contains about 350,000 polygons from a geographical database.

Figure 8.8
Virtual display lists: A stabilized platform-deployment station. See insert for color representation.

8.5.5 Ray Tracing

Each pixel node contains a copy of the display list for the scene to be rendered. Ray trees are traced in parallel by the pixel nodes, with each node generating ray trees for pixel sampling points of its unique set of interleaved pixels. If the size of the display list exceeds the local pixel-node memory, the display list is paged from the host computer, using the parallel page-faulting method described earlier. The active server is used to service display list page faults and texture map virtual shared memory requests, respectively. The pipeline is used to compute bounding volumes, tessellate geometric primitives, and transform the display list before rendering begins. The floating-point capability of each node is exercised to its maximum for the ray-object intersection tests. Antialiasing is performed by stochastic sampling in multiple passes.

Illustrations: The parallel paging scheme is employed for *virtual display lists* in the ray-tracing software implemented on the Pixel Machine. Figure 8.8 shows a ray-traced image with 17,000 polygons. Each polygon uses 100 bytes, giving a database size of 1.7 Mbytes, substantially more than can fit in one pixel node's local memory. Figure 8.9 also shows a ray-traced image generated using virtual display lists. This scene contains over 50,000 polygons, exhibits area-light sources, and is antialiased at 16 samples per pixel.

Figure 8.10 shows a ray-traced image that uses *virtual texture maps*. There are 13 virtual texture maps requiring a total of 4 Mbytes of texture map data. The scene also contains approximately 2000 polygons.

Figure 8.9
Virtual display lists: A tea room. See insert for color representation.

Figure 8.10
Virtual texture maps: A museum room. See insert for color representation.

Our first objective was to see if the performance would improve linearly with increases in the number of processing nodes for the case where the object database (Figure 8.11) resides entirely in a node's memory. In these tests, the database size is kept constant while the number of pixel nodes in a system is increased. The results of this test are plotted in Figure 8.12. As can be seen from the dotted graph, our objective of linear improvement was met, and we have had similar experiences with many other object databases. In addition, the actual rendering times for the image are plotted (solid graph) and they are two to three orders of magnitude faster than on typical workstations.

Our second objective was to examine what happens to the performance when the parallel paging is used for virtual display lists. This test was run keeping the number of pixel nodes in the system constant, while increasing the number of objects in a scene. It can be seen from the graph in Figure 8.13 that paging begins at about 5000 objects in the display list. The performance degrades exponentially when paging begins, but it again becomes linear above 10,000 objects. The dotted graph labeled "requested" shows the number of page faults generated by all the nodes. Because many nodes request the same page at the same time, the dotted graph labeled "serviced"

Figure 8.11
Ray-tracing performance: A sphereflake. See insert for color representation.

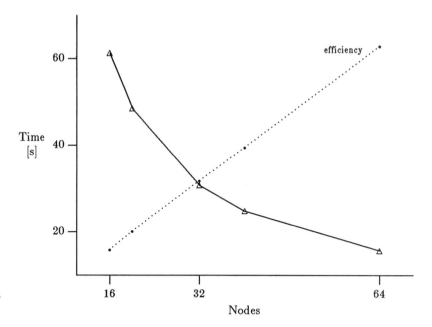

Figure 8.12
A plot of Sphereflake execution
times for 512^2 images.

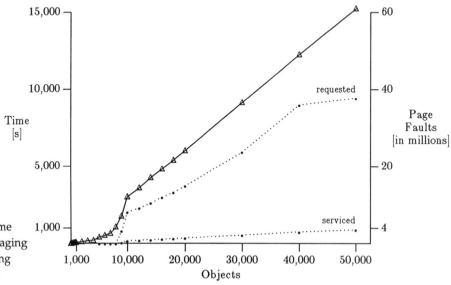

Figure 8.13
Scene complexity versus time
(solid curve) and parallel paging
(dotted curve) for ray tracing
using virtual data memory.

shows that only about one-tenth of the generated page faults had to be serviced. The speed of the algorithm when paging a display list is about five times slower than when all of a display list is in the pixel-node memory.

8.5.6 Volume Rendering

Rays are marched [10] in parallel by the pixel nodes inside volumetric data. Each node processes its set of interleaved pixels in the image. At each pixel, a ray is cast into the volume being rendered and ray-marching incrementally steps along the direction of the ray, sampling the signal inside the volume. The sampled values of a ray are then converted into image intensity according to the application: Thresholding, finding maximum, translucency accumulation, or integration can be selected. The volume is stored on the host computer, with each pixel node requesting voxel packets that contain voxels along the path of a marching ray. This procedure is accomplished using virtual shared memory via the active server. The pipeline is not utilized in this mapping. Antialiasing is accomplished by sampling very finely along each ray and by interpolating voxel values adjacent to an intersection point.

Illustrations. Figure 8.14 shows two volume renderings of a nuclear magnetic resonance (NMR) angiography study that uses *virtual vol-*

Figure 8.14
Virtual volumes: (a) A Sagittal View of NMR Data ($256 \times 256 \times 60$ voxels) and (b) A Transverse View of NMR Data ($256 \times 256 \times 160$ voxels).

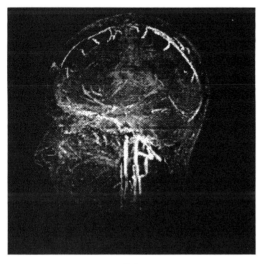

(a)

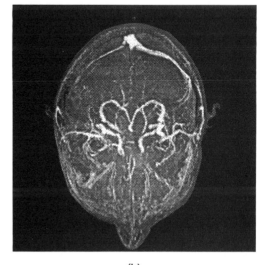

(b)

umes. The size of the data in Figure 8.14(a) is $256 \times 256 \times 60$ voxels (3.84 Mbytes), and the size of the data in Figure 8.14(b) is $256 \times 256 \times 160$ voxels (10.24 Mbytes).

8.5.7 Image Processing

An image is processed by the pixel nodes in parallel, with each pixel node computing its set of interleaved pixels. If the image is too large to fit in the local pixel-node memory, it can be distributed over the collective memory of all the nodes in contiguous block fashion and redistributed into interleaved format for a final display using the interleave/de-interleave strategy. The pipeline can be used for run-length decoding and other sequential image functions as an image is being sent to the pixel nodes.

Illustrations: Adaptive histogram equalization [11] is an image processing technique used to enhance contrast. At each pixel a histogram of the intensities of the neighboring pixels, called a *contextual region*, is examined. The histogram is equalized over the full intensity range, and a new value is assigned to a pixel based on where its intensity falls relative to the neighboring values.

Figure 8.15(a) shows four original 256 CAT scan images; Figures 8.15(b), 8.15(c), and 8.15(d) illustrate the AHE algorithm applied to the four images at each pixel with contextual regions of 63^2, 31^2, and 15^2 pixels, respectively. Figure 8.16 shows plots of the execution times for the adaptive histogram equalization on both 256^2 and 512^2 images using various region sizes.

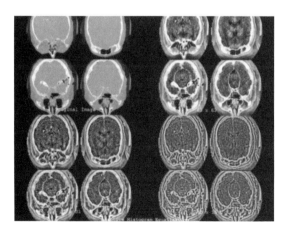

Figure 8.15
Image processing: Adaptive histogram equalization. (a) Original image, (b) 63^2, (c) 31^2, and (d) 15^2 pixel regions. See insert for color representation.

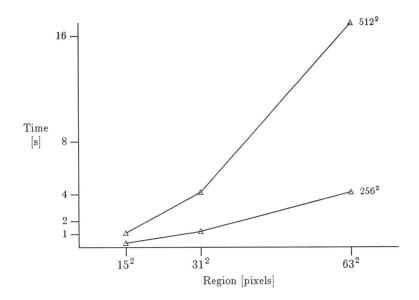

Figure 8.16
AHE execution times for 256^2
and 512^2 pixel images.

8.5.8 Fractal Functions

Fractal geometry is a branch of mathematics used to describe self-similar structures of fractional dimension [12]. The Julia set (Figure 8.17) is a class of fractal in the complex plane: A generating function is evaluated for discrete points in a rectangular region of the complex plane until the function diverges or a specified number of iterations is reached. The behavior of this function can be visualized by mapping

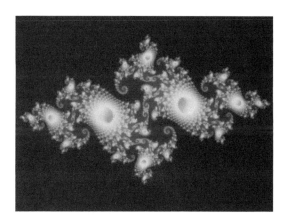

Figure 8.17
Fractal functions: A Julia set.
See insert for color
representation.

the complex plane onto a raster image. The color at each pixel is determined by the number of iterations for diverging points, and by the function value for converging points. This algorithm has been implemented solely in the pixel nodes: Each node computes the Julia set only at the complex points corresponding to the pixels in its portion of the frame-buffer.

8.5.9 Hypertexture Rendering

The architecture of the Pixel Machine allows new techniques to be developed for image computing that were previously not possible to consider. An example of such a technique is *hypertexture* [13]. This is a novel shape-modeling method, where the distinction between shape and texture is blurred and texture is used as a means for shape definition, with enormous computational requirements. This method produces images of richness and complexity not typically associated with computer imagery (Figure 8.18). Hypertexture also employs the ray-marching algorithm. However, there are only trivial data requirements. Instead of sampling an empirically obtained volume array, we are sampling a procedurally generated volume. This means we do not require the resources of the host computer to obtain data. Generation of 3D hypertextures, 2D solid textures, and fractal functions are similar: These techniques are procedural and trivially parallelizable, and they typically have linear speedups.

Figure 8.18
Hypertexture renderings: (a) A furry donut, and (b) A tribble.

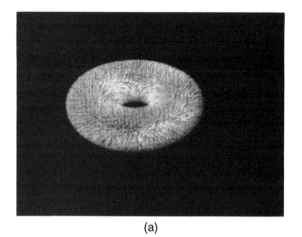

(a)

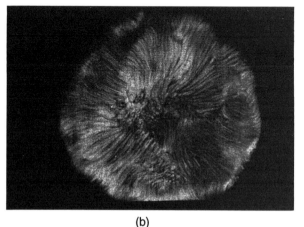

(b)

8.6 A MULTIMEDIA DEVELOPMENT SYSTEM

We have also used the computing power of the Pixel Machine to develop a research prototype of an interactive *multimedia* system. In addition to the graphics and image processing capabilities of the Pixel Machine described in this chapter, a multimedia system must be able to play digital audio and video data in real time. A pipeline is used for audio processing, while the other pipeline and the pixel nodes are used for video decoding:

- *Audio processing.* The upstream connection of a pipeline was modified to enable us to read and write digital audio data in a standard format [14]. The first node of a pipeline can read such data from CD and DAT players or digital audio receivers. The last node of a pipeline can write these data to DAT recorders and digital amplifiers. The data can be played back in real time, stored on the host, or played back from the host. The other pipe nodes (between the first and last nodes) currently only pass the data and digitally control the audio volume. In the future, real-time filtering, compression, and decompression algorithms can be added.

- *Video processing.* Compressed video sequences are read from secondary storage and decoded in the Pixel Machine in real time. A number of decoding techniques, ranging from simple block coding to discrete cosine transforms [15], have been tried balancing image quality, image size, frames/sec, and storage space. The compressed video data can be combined with compressed or uncompressed digital audio data which is separated and sent through the audio pipe to a digital amplifier for real-time synchronized sound.

These capabilities, for example, allow a user to interact via a front-end *hypertext* system with multimedia databases. Pointing to the word *zebra* will not only open a window with a short biography of zebras but will also show in another window a short video sequence of running zebras accompanied by the sound of their galloping. This certainly is a long way from just spinning triangles on the screen! We believe that visual computing of the future will be closely coupled with the video and audio media.

8.7 PARALLEL COMPUTING

Parallel computing is now being viewed (albeit with some skepticism) as a very promising route for dramatically decreasing execution times of computationally demanding problems that are intrinsically parallel in nature. The increasing presence of parallel computers in the marketplace attests to this, as do the reports in the computing literature [16] on order-of-magnitude speedups of various problems.

When the Pixel Machine was first developed, a number of people expressed interest in its use as a general-purpose computation engine. Since that time, a number of applications outside of the traditional image-computing domain have been written for the Pixel Machine. Two of these are described below.

8.7.1 A 3D Fast Fourier Transform

Dr. Henry McNutt from Biocryst, Inc. has implemented a three-dimensional fast fourier transform (3D FFT) in the pipe and pixel nodes. The 3D FFT is used to convert 3D data in *structure factor* form, empirically obtained from an x-ray diffractometer, to a 3D electron density map. The goal is to understand the molecular structure of purine nucleocide phosphorlayse (PNP). The 3D FFT is of size 128^3 with input and output data consisting of complex numbers represented by real and imaginary components as single-precision floating-point numbers; hence, the database is $128^3 \times 8$ bytes or 16 Mbytes. The database is distributed over the memory of the pixel nodes. In this implementation, the pipe nodes are used to transform the fastest changing index x of the FFT, with the pixel nodes transforming the other two indices y and z. Preliminary results show a performance of 40 Mflops with a 32-pixel-node Pixel Machine.

8.7.2 The Rendering Equation

Scott Hemphill at the Caltech Computer Graphics Research Group has an initial implementation of the rendering equation [17] running on the Pixel Machine. A Monte Carlo integration occurs over wavelengths in the visible spectrum (380 - 780 nm). The image is computed using the technique of successive refinement; at each pass, a ray is fired at every pixel. Each ray is traced from a jittered position in the neighborhood of a pixel using *importance sampling* with a triangular filter kernel. The ray is at a random wavelength in the visible spectrum; when it strikes a diffuse surface, it reflects in a random direction. Between 100 and 30,000 passes over the image are computed depending on the final image quality desired. Effects such as color bleeding are visible. The intention is to create images of diamonds with full internal reflections, caustics, chromatic aberration, and partial polarization properly simulated. Back-of-the-envelope computations show that one image which took 20 trillion flops to compute in 3 days on a Pixel Machine would have taken approximately 6 years to compute on a typical workstation.

8.8 SUMMARY AND CONCLUSIONS

We have described a parallel image computer designed for fast geometry and image computing. The computer contains a large distributed frame buffer which allows many computing elements, capable of floating-point operations, to access pixel-oriented data in parallel.

We have developed software for standard 3D polygonal graphics, 3D volume display, and ray tracing, all based on a common programming environment. We have also experimented with new approaches to visualization of 3D textures and to real-time playback of audio and video data. Others have used the machine as a general-purpose parallel computer, pursuing computationally demanding tasks on a practical time scale and at a practical cost.

To overcome the problems inherent to the architecture of the machine and its current implementation - particularly the limited amounts of program and data memories in each node - we resorted to using established software techniques found in traditional computers such as program overlays for instructions and virtual memory for data. This involved the development of the important concept of sharing resources between a parallel processor and a host computer. We also found a method to remove the restrictions of the interleaved frame-buffer design using interprocessor communication capabilities. To simplify the development of software for our scalable parallel architecture, we have developed a concept of physical and virtual nodes which makes the size of the machine transparent to the programmer.

It is our hope that people will not only make *faster* images using established techniques, but will also push the boundaries of what can be accomplished with image rendering by exploiting the potential of this new class of architecture for algorithm development.

The Pixel Machine was the first commercially available programmable *parallel* processor for geometry and image computing. We feel that the machine portends a paradigm shift in graphics architectures, which will become more apparent and widespread in the 1990s. We believe this trend will strengthen and produce a new wave of parallel graphics machines with unprecedented speed and flexibility due to their processing power and high communication and display bandwidths.

The objective is to allow visualization of large amounts of real-world and simulation data, requiring real time capabilities for interactive ray tracing, volume rendering, and polygon rendering of databases with unlimited complexity at the highest levels of realism and quality. Ultimately, the hardware should become a non-issue and the focus can shift to the optimal methods for man-machine interaction and the human interface. The Pixel Machine was an initial attempt to get us closer to this goal.

ACKNOWLEDGMENTS

We would like to thank Bill Ninke, Kicha Ganapathy, and Jim Boddie for providing a fertile environment that allowed the exchange of ideas between people involved in graphics, parallel processing, and digital signal processing. Leonard McMillan contributed major ideas to both the software and hardware architecture and should be identified as one of the principal architects of the system. Bob Farah should be credited with the design of the pipeline card and for handling enormous numbers of odds and ends. Marc Howard should be thanked for bringing to life very high quality, reliable video at 2 A.M. on a Friday night. Jennifer Inman should be thanked for writing a great deal of the pipe and pixel-node software.

We would also like to thank NASA for generating the image in Figure 8.8, Leonard McMillan for creating the image in Figure 8.9 and Kamran Manoocheri for the image in Figure 8.10. The airplane in Figure 8.7 is used courtesy of Alias Research, Inc.; the model was created by Steven Williams. The Sphereflake model in Figure 8.11 was designed by Eric Haines of 3D/Eye.

REFERENCES

[1] B.H. McCormick, T.A. DeFanti, and M.D. Brown, Visualization in Scientific Computing *ACM Comput. Graphics* **21**(6) (1987).

[2] R.N. Kershaw, et al., A Programmable Digital Signal Processor with 32-bit Floating Point Arithmetic, in *Proceedings of IEEE International Solid-State Circuits Conference*, pp. 92-93 (1985).

[3] C.L. Seitz, The Cosmic Cube, *Commun. ACM* **28**(1), 22-33 (1985).

[4] H. Fuchs, Distributing a Visible Surface Algorithm Over Multiple Processors, in *Proceedings of ACM 1977*, Seattle, WA, pp. 449-451 (1977).

[5] F.I. Parke, Simulation and Expected Performance Analysis of Multiple Processor Z-Buffer Systems, *ACM Comput. Graphics* **14**(3), 48-56 (1980).

[6] J.H. Clark, The Geometry Engine: A VLSI Geometry System for Graphics, *ACM Comput. Graphics*, **16**(3), 127-133 (1982).

[7] M. Potmesil and E.M. Hoffert, FRAMES: Software Tools for Modeling, Rendering and Animation of 3D Scenes, *ACM Comput. Graphics* **21**(4), 85-93 (1987).

[8] M. Dippé and J. Swensen, An Adaptive Subdivision Algorithm and Parallel Architecture for Realistic Image Synthesis, *ACM Comput. Graphics* **18**(3), 149-158 (1984).

[9] E.G. Coffman and P.J. Denning, *Operating Systems Theory*, Prentice-Hall, Engelwood Cliffs, New Jersey (1973).

[10] M. Levoy, Volume Rendering: Display of Surface from Volume Data, *IEEE Comput. Graphics and Appl.* **8**(3), 29-36 (1988).

[11] S.M. Pizer, et al, Adaptive Histogram Equalization and Its Variations, *Comput. Vision Graphics Image Processing* **39**(3), 355-368 (1987).

[12] B. Mandelbrot, *The Fractal Geometry of Nature*, W.H. Freeman, New York (1977).

[13] K. Perlin and E.M. Hoffert, Hypertexture, *ACM Comput. Graphics*, **23**(3), 253-262 (1989).

[14] Audio Engineering Society, Inc., AES Recommended Practice for Digital Audio Engineering - Serial Transmission Format for Linearly Represented Digital Audio Data, AES3-1985, ANSI S4.40-1985, *J. Audio Eng. Soc.* **33**(12), 979-984 (1985).

[15] W.K. Pratt, *Digital Image Processing*, John Wiley & Sons, New York (1978).

[16] G. Fox, et al., *Solving Problems on Concurrent Processors*, Vol. 1, Prentice-Hall, Engelwood Cliffs, New Jersey (1988).

[17] J.T. Kajiya, The Rendering Equation, *ACM Comput. Graphics* **20**(4), 143-150 (1986).

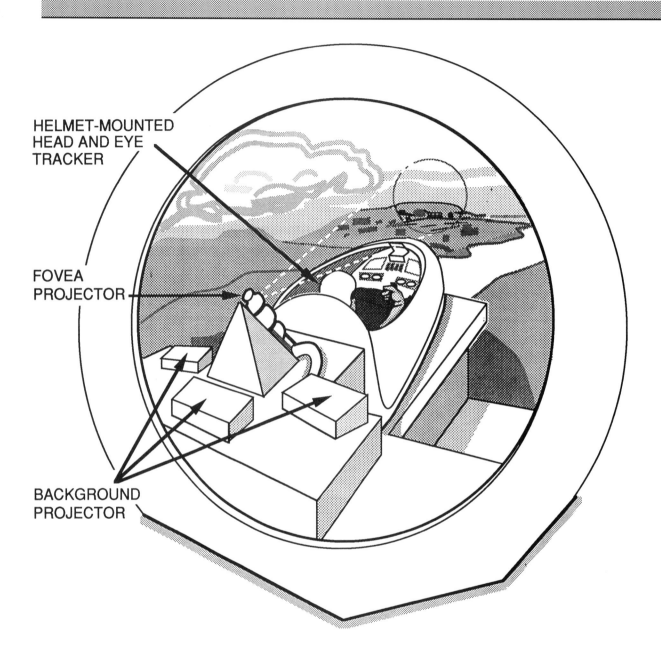

HELMET-MOUNTED
HEAD AND EYE
TRACKER

FOVEA
PROJECTOR

BACKGROUND
PROJECTOR

9

Brave New Virtual Worlds

David M. Weimer
Informations Systems Research Laboratory
AT&T Bell Laboratories
Holmdel, New Jersey

9.1 INTRODUCTION

These days the concept of "virtual reality" is everywhere: from science-fiction movies, to arcade games, to serious computer journals. Certainly computers of the future will allow us to enter lifelike, computer-generated realities. Already today you can gaze into a three-dimensional world limited only by the speed with which your computer can change the images in response to your eye and head motions. Some call this sense of actually being present within a new reality a "virtual reality" or "cyberspace." The degree to which cyberspace becomes undifferentiated from reality depends partly on the development of new ways for humans to physically interact with the computer's reality (such as goggles and gloves).

Virtual worlds, or virtual reality, is a technology continuing to draw a great deal of attention, much like artificial intelligence did in the seventies and robotics did in the early eighties. Virtual reality is defined as a basic technology that combines computer models, computer-generated images, multimedia, and real-time interaction. Virtual environments is another name that has been used for the same technology, but this term more appropriately describes specific applications, not the underlying technology. Environment simulators use a variety of techniques to give people the experience of being somewhere they are not (in a cockpit, on the moon, etc.).

There has been a recent surge of interest in virtual reality, not because it is a new concept but because many of the technical and cost barriers that have slowed progress are being pushed aside. The idea of virtual worlds has been around for some time. Aldous Huxley's book, *Brave New World*, published in 1932 [1], anticipated how virtual worlds of today, based on computers and transducers, would begin to sound, look, and feel like the real world. In it, he describes a futuristic entertainment medium that he called "An All-Super-Singing, Synthetic-Talking, Color, Stereoscopic Feely." In 1965, Ivan Sutherland presented a vision of virtual worlds that continues to inspire many [2]. His vision was influenced by the introduction of computer-driven display technology. He saw in it the potential for visualization beyond television or depiction of familiar scenes: "A display connected to a digital computer gives us a chance to gain familiarity with concepts not realizable in the physical world. It is a looking glass into a mathematical wonderland."

Three years later the first head-mounted display system was reported [3]. This helmet used half-silvered mirrors to merge views of the real environment with views displayed on miniature cathode ray tubes (CRTs). Both a mechanical and ultrasonic method were used to measure head motion, and these data were used to apply the corresponding viewing transformations to the graphical objects. Today's head-mounted displays use slightly more advanced components, but they are all based on the same concept. Subsequent to Sutherland's pioneering work, there have been many other efforts towards implementing virtual worlds. Some significant milestones are listed below.

- The Lincoln Wand: a hand-held wand with ultrasonic 6 degree of freedom (dof) tracking [4].

- A force feedback joystick was developed by Michael Knoll [5]; at about the same time, project GROPE-I was started [6].

- Sorcerer's Apprentice: a head-mounted display and wand [7].

- The 3SPACE[1] was developed at McDonnell Douglas.

---

[1]3SPACE is a registered trademark of Polhemus.

- Put-That-There: Richard Bolt combined speech recognition with 3D pointing [8].

- Myron Krueger coined the term *artificial reality* and published the first book on the subject [9]. His implementation, called Video Place, uses noninvasive approaches to tracking [10].

- A tracking glove was developed by Tom Zimmerman and Jaron Lanier [11].

- A virtual environment for teleoperation was developed at NASA [12].

- The Human Interfaces Technology Lab was formed at the University of Washington. It is directed by Tom Furness, who worked for over 20 years developing flight simulators for the Air Force.

This is just a small sample of the work that has contributed to virtual reality. While the technology is still not mature, it is now at a stage where more researchers can easily participate. A common theme drives all research in virtual reality: an ultimate display system that can simulate various phenomenon realistically. There is also another theme that exists, slightly different from the first: simulations that go beyond presenting alternative environments, and act in ways that augment human activity within real environments.

This chapter consists of four sections that examine the following aspects of virtual reality: basic device technology, applications, and research issues. Section 9.2 is a survey of enabling technologies that are associated with the creation of virtual worlds. It describes the history, the current state of the art, and the immediate future of these enabling technologies. Section 9.3 discusses a broad range of applications, broken down into three different categories of work: shrinking time and space, visualization, and recreational applications. Many of the applications that are discussed are futuristic because most virtual environments are still research prototypes. Applications were chosen that fulfill one or more of the following criteria: it solves a real problem, it presents substantial technical challenges, it offers potential competitive advantages, or it provides a unique form of service or entertainment. Some applications are too far off to predict with any accuracy, but they are presented as a glimpse of what might be possible. The concluding section outlines the current research and development issues in virtual reality, specifically those related to basic device technology and systems integration.

9.2 DEVICES AND ENABLING TECHNOLOGY

This section will describe enabling technologies for virtual reality. Virtual reality is a concept that incorporates more than an array of IO devices. It will continue to rely on technology in other areas: computer hardware, algorithms, and networking. This survey will attempt to reach into these areas and will identify many of the pieces that are likely to impact research and development efforts.

9.2.1 Displays

Output devices have advanced at a somewhat faster pace than devices for input, perhaps due to the origins of virtual reality in flight simulation. In flight simulation, visual feedback to a pilot has been the primary focus of innovation as opposed to input techniques because the controls must be those of a real aircraft. For this reason, flight simulation is a good starting point for discussing virtual reality display technology.

Exploiting Visual Dominance

Many present simulators rely only on computer-generated scenes. Realism is a crucial factor, and proper visual feedback is often the most important aspect of creating effective simulations. "Visual dominance" is a well-known aspect of human perception in which visual cues override the other senses. This trait is often exploited by displaying scenes that fill as much of a person's field of view as possible. Some techniques for providing wide fields of view include the following: multiple CRTs, projection screens/domes, variable resolution displays, and head-mounted displays.

Multiple CRTs were the first attempt at providing wide-angle viewing. The main advantage was use of standard, commercial displays rather than expensive customized displays. There are, however, some obvious problems with the approach: (a) limits on scaling overall display to larger sizes, (b) duplication of image generation hardware for each display CRT, (c) a limited field of view (for example, making simulation of glass canopy environments of jet fighters difficult), and (d) inability of square CRTs to represent a wide, unbroken view.

Projection techniques offered some advantages over CRT arrays. They provided a way of enlarging a single image to cover a larger area, and they provide a slightly brighter image. However, because they enlarge both the image and the individual picture elements, effectively reducing the display resolution, simulators often require multiple projection systems (e.g., using multiple flat screens or a single projection dome). Duplication of expensive imaging equipment places a limit on the number of high-resolution projections that can be used.

Variable-resolution displays offer an alternative method for increasing resolution. It is well known that human vision is spatially coarse in the periphery, but has the highest resolution within the fovial region. A variable-resolution display takes advantage of this property through the use of an eye tracking system. The display system tracks the pilot's gaze and inserts a separate high-resolution image within the area seen by the fovia. The high-resolution image is a circular area slightly larger than the diameter seen by the fovia and is blended with a coarse-resolution background image [13]. A dome projection system with a gaze-directed fovial projector was conceived at Singer's Link Flight Simulation Division, shown in Figure 9.1. While it introduces added hardware costs and more complexity to the rendering algorithms, variable resolution has proven to be effective in increasing display resolution without increasing the rendering time. This improved level of performance has also been demonstrated in a system for displaying volumetric data [14].

Head-mounted displays are the latest approach that is being considered for flight simulators. Some believe this approach can reduce costs by incorporating all of the display equipment, optics, and eye tracking in the helmet. It turns out that Sutherland's head-mounted display established important concepts that continue to guide research in display technology. Today there are many applications outside of flight simulation that are exploring the potential of head-mounted displays. For this reason, they deserve a close look.

Head-Mounted Displays

Head-mounted displays are currently the most effective way of visually surrounding someone with computer synthesized environments. Because the CRTs are part of a headset or helmet, the eyes can see the faces of the CRTs wherever the head is positioned. The head position and orientation is tracked continuously, and the simulated environment is viewed in a way that is consistent with the head motion. If a simulated object is placed in a fixed position, a person wearing a head-mounted display can move around the object and see it from any side. The head-mounted displays designed by NASA and VPL

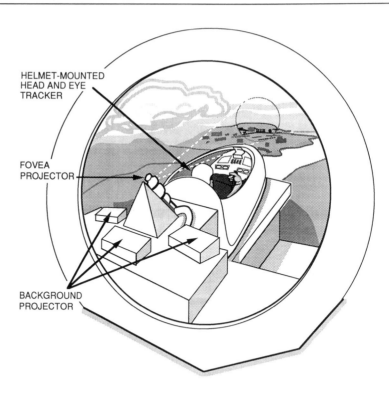

HELMET-MOUNTED
HEAD AND EYE
TRACKER

FOVEA
PROJECTOR

BACKGROUND
PROJECTOR

Figure 9.1
Flight simulator with a
projection dome and a
variable-resolution gaze-directed
display.

use this simple idea and let the user see only what is presented on the CRTs. These systems use a pair of color liquid crystal displays that are mounted in front of the eyes. Lenses are placed between the displays and the eyes to achieve both a magnification of the images and a more distant focal plane. For each eye in the VPL display the field of view is $80°$ horizontally and $60°$ vertically. It is important in any head-mounted display system that the eyes are not strained by having the focal plane be too close. In a head-mounted display for flight simulation, the optics may also include a layer that collimates the light rays. The light rays are made parallel to simulate views of distant terrain that are typically seen during actual flight.

The system developed by Sutherland used a slightly different arrangement of displays and optics. In his design, computer-generated images were merged with views of the existing surroundings. This was accomplished by using half-silvered mirrors, *beam splitters*. The technique is illustrated in Figure 9.2. This concept has been used in a display system that was developed for the Air Force by CAE Electronics Ltd. in Montreal [15]. This helmet may be the most advanced

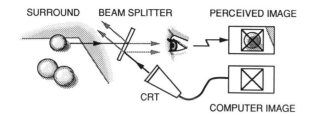

SURROUND   BEAM SPLITTER   PERCEIVED IMAGE

CRT

COMPUTER IMAGE

Figure 9.2
A technique for combining images using beam splitters.

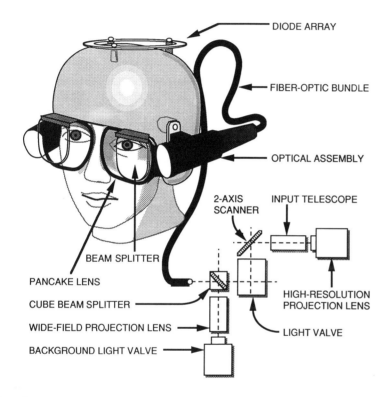

DIODE ARRAY

FIBER-OPTIC BUNDLE

OPTICAL ASSEMBLY

2-AXIS SCANNER   INPUT TELESCOPE

BEAM SPLITTER

PANCAKE LENS

CUBE BEAM SPLITTER

WIDE-FIELD PROJECTION LENS

BACKGROUND LIGHT VALVE

HIGH-RESOLUTION PROJECTION LENS

LIGHT VALVE

Figure 9.3
A diagram of the CAE head-mounted display showing the imaging for a single eye.

head-mounted display yet developed. It combines two beam splitters for each eye: one that merges the surroundings with synthetic images, and one that combines two different computer-generated images (see Figure 9.3). One of the computer-generated images serves as a lower-resolution background image that is slaved to the head motion. The other is a high-resolution inset that is slaved to both the head and eye motion. This implements a head-mounted version of a gaze-directed variable-resolution display. Background images cover the same field of view as the VPL display, while high-resolution insets cover 25° horizontally and 19° vertically. A fiber bundle is used to relay the computer-generated images to the optical assembly on the helmet. One major problem with this and other head-mounted displays is their weight; the CAE helmet weighs a hefty 5 pounds.

At the opposite end of the spectrum is a display that is based on much simpler technology. For those who only need bilevel images with monocular viewing, the Private Eye[2] may be an adequate display device. This system uses a single scanline of 280 LEDs that is optomechanically shifted 720 times to display an entire image, with the entire image updated 30 times a second. This resolution is capable of displaying a screen of text that is 80 columns by 25 lines. It weighs approximately 2.25 ounces, and the unit's dimensions are relatively small: 1.2 inches high × 1.3 inches deep × 3.5 incues long. The image appears as a computer screen that is suspended 2 feet in front of the viewer. Because only one eye is covered, the real environment can still be seen. This may be the least invasive form of head-mounted display on the market today and is the most practical alternative for office settings.

One of the major problems with head-mounted displays is their vulnerability to latency. If the scene update rate is 60 Hz, it takes 16 msec to completely change the display. Some experts maintain that, for rapid head motion, anything above 5-8 msec will be noticeable and may cause motion sickness.

Displaying Stereo Images

Most of the preceding examples are stereoscopic systems, which present different views to each eye, introducing lateral parallax. There are two basic categories into which the different techniques can be placed: those based purely on optics and those based on optomechanical devices. The techniques also fall into two categories of presentation style: those that require special glasses and those capable of providing auto-stereo. Figure 9.4 shows a diagram of existing techniques for stereo imaging.

The oldest form of stereo viewing used separate image pairs and displayed them to each eye along separate optical paths. Early photographic stereograms were produced by using (a) a pair of cameras and (b) special viewers that delivered left and right images separately to the appropriate eye. Head-mounted displays operate on the same principle: Each eye has a separate CRT. Techniques that use separate

---

[2]Private Eye is a registered trademark of Reflection Technology.

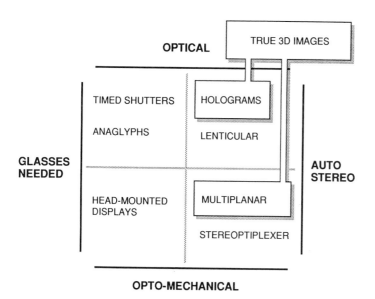

Figure 9.4
A diagram of the existing techniques for displaying stereo images.

optical paths, moving apertures, or moving reflective surfaces will be referred to here as *optomechanical techniques*. Other techniques that rely on physical properties of light (such as wavelength, polarity, or phase) will be referred to as *purely optical approaches*.

Lenticular Stereo: Lenticular auto-stereo has been used for a very long time to produce three-dimensional postcards and pictures that appear to be animated as the picture is rotated from side to side. Lenticules are small cylindrical lenses that are superimposed over a series of vertical image stripes. The left and right image stipes are interleaved and are viewed through the lenticules. When properly registered, the lenticules provide separate optical paths for two image strips, one for the right eye and the other for the left eye (see Figure 9.5). This technique only works for a limited range of viewing positions.

A device known as a *stereoptiplexer* is an extension of a technique developed by Ives that used lenticules and projector arrays [16]. The stereoptiplexer uses a rotating slit aperture to project omnidirectional stereo images without glasses. Collender [17,18] developed a series of stereoptiplexers that used standard movie film, a rotating optomechanical scanner, and a retroreflecting projection screen with vertical scattering. He also developed a technique for three-dimensional television without glasses, but so far these systems have not been commercially developed.

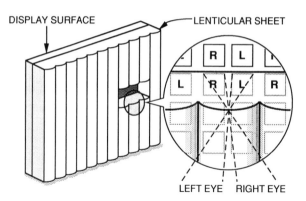

DISPLAY SURFACE   LENTICULAR SHEET

LEFT EYE   RIGHT EYE

Figure 9.5
A diagram showing the structure of a display for lenticular auto-stereo.

Anaglyphic Stereo: Anaglyphs are image pairs that are projected through the same optical paths and that rely on optical filters (e.g., special glasses) to separate left and right eye views. The earliest form of anaglyphic stereo used monochrome photographs (Rollman in 1853) and at the turn of the century work began on anaglyphic motion pictures [19]. In 1945, Dudley [19] developed a method for directly recording polychromatic anaglyphs on ordinary color photographic film. Again, complementary filters, each with a wide waveband, were used to isolate the left and right views, thus allowing each eye to see full-color stereo images.

Polarized anaglyphs use light polarization rather than wavelength to isolate left and right views. For each eye a pair of plane polarizing filters are used. If two polarizing layers have their polarizing axes parallel, they can transmit light at roughly 50% efficiency. When one polarizing axis is rotated by 90° in the image plane, all light is blocked. Polarization can be used for viewing motion pictures, provided that the projection screen does not depolarize the light. Two commercial systems have been developed that use polarization to view stereo images on regular CRTs. These systems use LCD shutters that switch polarization 60-120 times per second. The shutters are synchronized to the display of alternating left and right views on a CRT. A predecessor of LCD shutters [20] was a device called PLZT glasses, made of lanthanum zirconate titanate wafers. These were very expensive and had a short lifetime.

Multiplanar: A multiplanar technique generates a spatial image by sweeping (at a cyclic frequency higher than can be visually perceived) a time-varying two-dimensional image through space. An early prototype of a multiplanar technique was the verifocal mirror [21], using a mirror membrane (aluminized Mylar) mounted on a vibrating loud speaker and reflecting the images of an adjacent CRT. The image on the CRT was synchronized to the frequency of the vibrating membrane. The apparent depth of the image was roughly 80 times the distance traveled by a flexible reflector at its center, which is much better than the factor of two that can be achieved with a flat mirror.

In the OmniView system (developed by Texas Instruments), a transparent disk is rotated within a plastic dome. The disk is tilted with respect to the horizontal plane and spins at 600 rpm. At the same time a low-power laser, modulated at 10,000 times per second, illuminates the disk by sweeping the beam in a two-dimensional scanning pattern. With red, green, and blue lasers the system has the potential to display full-color images. Both the verifocal mirror and the OmniView displays create auto-stereo with appropriate left and right eye parallax both horizontally and vertically. Unlike the verifocal mirror, the OmniView displays true three dimensional images, with a omnidirectional viewing range (360° horizontally and 180° vertically).

Holography: Holograms are recorded on high-resolution photographic plates, storing the spatial variation of the light intensity and its phase at the plate by illuminating the scene with two coherent waves of light. These two coherent waves beat together to form a standing interference pattern, which can be encoded as an intensity image. Reconstruction is achieved by passing an incident monochromatic beam through the holographic film. The fringe pattern on the film diffracts a portion of the beam and forms the original incident wave [22].

The Ultimate Display

In the ultimate display, objects would appear to exist in three dimensions and could be manipulated directly with the hands. This defines the principal objective, independent of any particular technology used for its implementation. It is important to evaluate whether or not a technology, as it advances, will eventually provide support for the basic objectives. Ultimately, camera images will be capable of tracking human motion for input to an environment simulator, but most of the current work in stereo display will not allow the hands to reach in.

The only exception is the head-mounted display. As invasive as it is, it is the only display technique that can allow someone to reach into a simulated three-dimensional scene as though both the hands and the scene existed in the same space. Anything else requires the use of a virtual hand that is controlled from a space that is visually separated from the real hand.

Audio Feedback

In multimedia interfaces and virtual environments, the use of audio has been proposed as a means of enhancing human perception. There are two basic categories of audio stimulus: variations in the temporal domain and variations in intensity level. Variations in the temporal domain can produce the following types of perception: frequency as tones, frequency as density, or separated tones as cadence. The temporal domain also accounts for some forms of spatial perception, (eg., binaural cues, Doppler shifts, and reverberation). Intensity variations can be perceived as loudness, or they can be perceived as spatial cues, (eg., binaural cues, relative proximity, and rate of approach).

Studies have shown that differential sensitivity to steps in frequency (pure tones) is very acute. For moderate loudness levels, the difference threshold is about 1 to 3 Hz over a range of 62.5 to 2000 Hz [23]. Differential sensitivity to intensity (loudness) is sensed with slightly lower acuity than changes in pure tone, but still provides a high-bandwidth channel. Small differences in intensity can be detected for pure tones around 4000 Hz: 0.2 to 1 dB over a range of 20 to 100 dB SL[3] [24]. This provides roughly 282 steps from 0 to 140 dB SL, a differential sensitivity that is more acute than the eye's ability to discriminate between intensity levels. There are many other features of sound that can be used, which combine aspects of frequency and intensity. For example, *timbre* describes a tonal quality of sound that is determined by a group of wave forms and their harmonics. People can readily discriminate between variations in timbre, which has been likened to visual perception of color [25].

There are a few research efforts underway in synthetic audio environments. One example of a directional audio system based on speaker arrays was developed for cockpit applications [26]. Another system, developed at NASA-Ames, used real-time digital filtering to generate three-dimensional sound cues over headphones [27]. Bin-

---

[3] The *sensation level*, abbreviated SL, is the number of levels above an individual's threshold.

aural cues rely on the effects of sound waves from a given source arriving at the two ears at different times and at different intensities. The shape of the outer ear also provides monaural cues for locating sound sources. To avoid modeling these effects in detail, the simplest approach is to measure the response to an audio impulse signal (a short spike of high amplitude) through tiny microphones placed in the ear canals. Samples of the impulse response can be taken for many positions around the subject and can be used to construct head-related transfer functions (HRTFs) [28]. The HRTFs are used to convolve a digital sound source, recreating the original temporal and intensity variations that give binaural cues. Special-purpose hardware is required for real-time convolution, while head tracking is used to select HRTFs. In this manner a sound source can be fixed in space, regardless of head position. Intensity and timing do not vary between the two ears when the source lies on the azimuth-equal-to-zero plane. This results in ambiguities where sound can be perceived as coming from the opposite direction. There are also potential ambiguities between points located along areas known as *cones of confusion*.

There are many applications in scientific visualization in which audio can be used to enhance perception of salient features. Audio has been proposed as a useful modality for perceiving turbulence [29]. It has also been used in conjunction with two-dimensional graphics to visualize empirical data of high dimensionality [30]. The auditory channel also provides some capacity for loading. For example, people have no trouble listening to music while carrying on conversations. Hence, a combination of synthetic speech and musical audio may offer parallel information channels that avoid sensory overload. Auditory patterns have also been used as icons, also referred to as *earcons* [25].

Haptic Displays

Like the sense of hearing, the haptic senses can discriminate some things better than the visual system. Haptic perception describes an active exploratory process of feeling and doing that combines sensations of the skin and skeletal muscles [31]. The cutaneous senses are those of the skin and can sense a number of tactile stimuli such as contact, texture, vibration, friction, temperature, and so on. *Proprioception* refers to the sensations that are felt internally by the muscles, joints, and tendons. These receptors are used for sensing physical properties such as resistive forces, mass, stability, shape, and so on. One of the challenges of virtual reality will be to deliver appropriate stimuli to the haptic system, either for the enhancement of realism or for the encoding of information. There are a few existing systems that use haptic display, some favor the cutaneous senses (tactile display), while others favor proprioceptors (force display).

Force Display: Force display devices are usually constructed from actuators paired with potentiometers, and they automatically double as input devices. One of the earliest examples of a haptic IO device was a three-dimensional joystick built by Noll [5]. It used potentiometers and two-phase induction motors to apply forces within a 10-inch cubical space. Without applying any forces, the device had a high degree of inertia and required a control strategy that provided power assistance. At about the same time a two-dimensional haptic IO device was developed by Batter and Brooks [6], consisting of a knot on a plate that could be moved within a 2-inch square at 12 updates per second. The device was used to examine continuous force fields. A more recent implementation of a two-dimensional force-reflecting joystick was developed at MIT by Max Behensky and Doug Milliken and was capable of 1000 updates per second. It has been used for empirical studies of discrimination of textures (i.e., relative roughness of a surface), as well as some physical systems based on springs, dampers, and mass [32].

Another early device that predates all of these is the first master-slave teleoperator developed by Goertz in 1945 at Argonne National Laboratories. This device was used for remote handling of radioactive material. Later this device was redesigned in order to replace the direct mechanical linkages with electrical servomechanisms [33]. The most recent version, the ANL-3, was a six-degree-of-freedom force-reflecting master [34]. Two surplus masters were donated to the University of North Carolina at Chapel Hill, where they were used as part of project GROPE. Early experiments used it to manipulate a pair of virtual tongs within a simple blocks world [35]. More recently, it has been used in a series of empirical studies for evaluating its effectiveness in a molecular docking task [36]. Project GROPE will be discussed in greater detail in Section 9.3.2.

All of the haptic IO devices mentioned so far can deliver a sensation of force delivered from a single point source, much like what can be sensed by poking around with a pencil. They all tend to favor the proprioceptive senses throughout the hand, arm, and upper body. Minsky's study showed that computer-generated forces could be used to convey tactile information, and subjects had no difficulty in discriminating between varying degrees of roughness.

There are other examples of force reflectance that attempt to deliver more than a single point of force reflectance. Noll suggested that separate forces could be delivered to each digit, allowing the hands to feel shapes in the way they are accustomed. This approach is used in a nine-degree-of-freedom device that was recently developed at the University of Tsukuba in Japan [37]. Iwata's force reflecting master is a small desktop unit that delivers separate forces to the palm/back of the hand as well as the thumb and index finger. The rest of the fingers receive a single force as a group. A single force can be applied in opposition to the flexion of all three finger joints as a whole. The thumb also receives a single resistive force that opposes abduction and adduction. These three actuators ride on a platform, and the remaining three degrees-of-force and three degrees-of-torque are applied to the plate, delivering additional forces to the palm and fingers. Other exoskeletal devices for the arms, torso, and hand are being developed for teleoperation of anthropomorphic robot manipulators. This application will be discussed in Section 9.3.1. One of the problems with force reflectance using exoskeletons is the degree to which they are invasive. Because they require mechanical systems, they are currently very bulky and can be very heavy. A nice feature of Iwata's desktop master is that it does not rest exclusively on the hand.

Tactile Display: The oldest examples of tactile display are those used for the blind, namely documents in the form of braille. Contact with the skin can be felt long before enough force is generated to stimulate the proprioceptors. The skin can also discriminate between patterns over an area of contact (i.e., distinguishing edges or corners from surfaces, and rough or grooved areas from smooth ones). Current technology for delivering tactile stimulus to the hands and fingers may be even more difficult than force reflectance, depending on which sensations are targeted.

One simple approach to delivering contact information is through vibrational stimulus. This method has been explored as a technique that can be incorporated with hand-tracking gloves. Vibration can be an effective substitute for pressure to the skin, but there are problems in making small vibrating elements. They also tend to produce a tingling effect, which may limit their application to arrays with very coarse resolution. This method has been proposed as an adjunct to force reflectance as well as a substitute for it. Research in the area of micro-mechanical structures may be the area from which techniques emerge which can deliver pressure stimuli to the skin, at a resolution that matches the resolution of receptors.

9.2.2   Input Devices

This survey now turns to the input side, to examine the devices and enabling technologies that translate the actions of humans into events that control virtual worlds.

Three-dimensional tracking is an important research topic in a variety of fields. Here, the discussion will be limited to those methods that are aimed at tracking human motion.

Ultrasonic Devices

Ultrasonic tracking uses time delay of sound signals and can sense rigid-body motions (positions and orientations) in three-space using triangulation. Sutherland's ultrasonic tracking system used continuous-wave ultrasonic signals that transmitted three different frequencies from a headset to four receivers mounted in the ceiling. Phase changes could be measured along 12 paths and were used to establish a three-dimensional position [3]. Another technique, *pulsed* ultrasound, was used for the Lincoln Wand [4]. This technique may be more susceptible to interference from ambient sound, but can sense delay over far more than a single wavelength, eliminating the need for counters that count complete changes in phase.

Electromagnetic Devices

The Polhemus Navigational Sciences Division of the McDonnell Douglas Corporation developed a tracking system based on electromagnetic pulses [38]. The 3SPACE tracker uses a source with three orthogonal transmitter antennas. Each transmits a pulse in sequence to three orthogonal receiver antennas, at a rate of 60 times per second. Each pulse induces a different current in the sensor antennas, which varies according to distance and relative orientation. Each series of three pulses produces nine currents which are used to determine three coordinate values and three degrees of orientation. The sensor can operate within a meters radius of the source and can have positional accuracy within ≈ .03 inches within a 20-inch radius.

Optical Tracking Devices

Two forms of optical tracking exist for tracking single rigid-body motion: those that are active and those that are passive.

Traditional active approaches use images from cameras located in the room which track a target with active light-emitting elements. SELSPOT is a commercial system that uses two camera-like fixed sensors that detected the three-dimensional position of up to 30 light-emitting diodes using stereo triangulation methods [39]. If the target lights are fixed in a known spatial relationship, the system can also measure the orientation. Another system, Twinkle Box, used four light-sensing devices to detect light from sequentially blinked LEDs [40]. Like SELSPOT, it can determine orientation using multiple LEDs in a known spatial relationship, but its sampling rate was only 61 lights per second versus SELSPOT's rate of 30 lights at 315 samples per second.

A passive system is one that uses no special light-emitting elements, just the light that is transmitted by a scene. These are the least invasive forms of tracking, but they are also the most difficult to implement. For some applications, controlled lighting can be used to simplify the tracking problem. Another way to simplify the problem is to adopt an outward-looking approach. Self-Tracker was an outward looking tracker proposed by Bishop [41]. This device used multiple one-dimensional light-sensing arrays, built into an image processing chip. Each array would look out at the world and sense relative motion by detecting one-dimensional shifts in texture. The aim was to provide simultaneous tracking of several users over larger areas. A more recent outward-looking system uses three head-mounted cameras along with an array of LEDs mounted on the ceiling [42].

All of the vision-based tracking methods above attempt to track single or multiple rigid-body motions. There are some efforts underway that attempt to use computer vision to track nonrigid bodies, in a manner similar to that of the DataGlove. Myron Krueger developed Video Place to demonstrate the use of noninvasive techniques for creating synthetic environments [10]. One of the most interesting components of Video Place is its use of video as input. A single camera view and algorithms for feature extraction are used to implement a two-dimensional interface that recognizes simple gestures. A similar capability has been implemented by Segen [43].

The current state-of-the-art of computer vision is not capable of accurately tracking a hand as a nonrigid body in three dimensions. For this task, more invasive techniques are still required.

Hand Trackers

Hand trackers combine positional trackers (such as the Polhemus) with sensors that measure the flexion/extension of the finger joints. There are various techniques for sensing articulated motion of either the hand or the entire body. Those based on exoskeletons (e.g., Exos Hand Master) tend to be the most accurate, but at the same time can be more bulky. The Exos master measures the finger joints using Hall-effect sensors that are built into an exoskeleton [44]. The exoskeleton rides on the back of the hand and straps onto each finger link with Velcro strips. Exoskeletons may ultimately have an advantage when force or tactile display is also desired.

A competing technology is the glove developed by Tom Zimmerman and Jaron Lanier, marketed as the VPL DataGlove. It is less bulky than an exoskeleton because it uses fiber-optic loops (specially treated to have bend-dependent transmittance) to sense flexion. By treating only localized regions of the fibers, the effects of bending can be localized to a specific area, which helps reduce the coupling of joint motions. This type of device was used by NASA for virtual environment simulations [12] and has since been used in a number of other systems. It has been used for controlling robotic hands [45,46] as well as virtual environments for computer-aided design [47]. Another tracking glove, based on simpler technology, was developed by Kramer and Leifer at Stanford University. The TalkingGlove[4] has been used to implement a fingerspelling input device that generates synthetic speech for nonvocal people [48].

Eye Trackers

The objective of eye tracking is to detect the subject's line of sight in relationship to a display or surround. Some eye tracking systems are mounted on a headset, whereas others operate from a fixed position just a few feet in front of the face. If the eye tracker is firmly mounted on the head, the results are more accurate and rotations can be measured over a larger range. Because this only measures the rotation

---

[4]TalkingGlove is a registered trademark of Virtex, Inc.

of the eye in the socket, a head-tracking device must also be used to compute the line of sight. If the eye tracker is not mounted on the head, a technique must be used to isolate the effects of head motion on the pupil center. This is accomplished by the "pupil-center, corneal-reflection distance" method developed commercially by Young and Sheena.

The effectiveness of eye tracking as an input device has been the subject of a variety of experiments. Eye tracking can be used to gauge the viewer's point of interest, in order to orchestrate the display of information by a computer [49]. Eye motion can also be used to move cursors on the screen, which is very useful for severely handicapped users. There is still some doubt as to how effective eye movement is when compared to other means of input. In the real world, objects cannot be manipulated by gaze, and therefore it is not a form of manipulation with which people are familiar. There is also a tendency for the eye to jump rapidly or *saccade*, which also detracts from its potential for controlling a cursor.

9.3   APPLICATIONS

The applications for virtual reality vary in a number of ways. Some focus primarily on a single user experiencing simulated environments and events using head-mounted displays. Here most of the costs are associated with the display. Others focus less on environments and more on interfaces and multiuser participation – more costs and attention to input. Virtual reality can be used as a communication medium, as a tool for augmenting human productivity, or as entertainment. This section describes several applications considered important for virtual reality. The applications have been partitioned into three separate themes: shrinking time and space, visualization, and recreation.

9.3.1   Shrinking Time and Space

One of the most obvious examples of shrinking time and space is the telephone. As communication systems become more advanced, adding video and other media, society will increasingly approach a level of communication known as *telepresence*. There are two separate areas of research that are represented by the term telepresence. We will start by describing telepresence as it relates to telecommunications. Here the purpose is to improve the quality of communications through multimedia to the point where people seem like they are in the same room, even when they are separated by great distances. This idea is eloquently expressed by Bob Lucky in his book *Silicon Dreams* [50]. His notion of the dial on the telephone labeled "presence" would be like a brightness and volume control for realism. "When the dial is turned fully *on*, the person materializes fully, so that I can talk face-to-face." The settings would range from *off* (no communication whatsoever) through varying levels of presence that include: teletype, then telephone, then fuzzy video-phone, then a range of picture quality towards full presence. Here the goal is to improve the quality communication between two or more people. Telecommunications may be one of the most important applications for virtual reality

The field of robotics has its own definition of telepresence. A telerobotics system would use telepresence to relay sensory data (images, sounds, forces, etc.) to human operators, making them feel as though they were physically present at a remote site. In some cases the purpose would be simple observation, and in others the purpose would be to operate a vehicle or a robotic manipulator, as illustrated in Figure 9.6. Telerobotics uses this type of remote sensing and remote control for executing manipulation tasks in hostile environments: toxic, under-water, space, mining, and so on. It might also be an environment that is not at human proportions, either microscopic or macroscopic. Having a human in the loop can be better than purely automated approaches, especially for tasks that are one of a kind and for those that require dexterity or intelligence.

Virtual reality relies on some of the same kinds of technology that telerobotics does. There is no better example of this than the ANL-3 remote master that was mentioned earlier. It has been used both in telerobotics and in experiments in scientific visualization. Some techniques developed for virtual reality can be useful in telerobotics applications, especially in some kinds of supervisory control [51], on-line image enhancement, on-line plan simulation [51], or time-delay compensation [52].

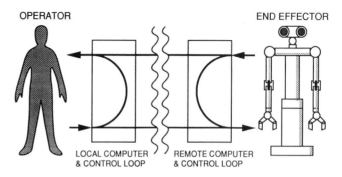

Figure 9.6
Conceptual diagram of telerobotics. (Adapted from a diagram in reference 51).

Virtual reality may eventually provide better tools for information delivery. Before the time of computer displays and cheap computer archival facilities, people like Vannevar Bush expressed concern for the lack of information appliances [53]. In his time the approach would have relied on microfilm and electromechanical devices. However, even with the computers and networks of today, technology has still not delivered the ideal information appliance.

Imagine what it would be like to have a computer library that you could access from home. It could be open 24 hours a day and would contain most of the information in the world. Unlike today's library, you never have to wait until a book is returned, because the original items are copied instead of checked out. The information might come in a large variety of formats. Since the earliest versions of *hypertext*, researchers have embraced the idea of electronic documents in an electronic library. When audio and video media are introduced into a document the format is referred to as *hypermedia*.

One extension to the computer library service would be the *virtual museum*. Virtual museums could containing vast collections of art and relics. Some of the interactive techniques that virtual reality attempts to provide today, when practical, will be the most essential tool for browsing through the museum's collections. While some items are two-dimensional (suitable for HDTV), many others are three-dimensional and require stereo viewing from arbitrary vantage points. If collections can be presented realistically, in a three-dimensional format, then virtual museums could include relics of any scale: whole ruins, ancient cities, and tall ships.

There are, however, many technological hurdles that stand in the way before such applications can be realized. Display technology, as it exists today, does not have the presentation quality that would be required by a virtual library or museum. A display would need a resolution equivalent to 300 dots per inch before it could begin to match the clarity of a textbook. There are also problems with collecting massive amounts of data to be stored in electronic form.

Text as a scanned image does not pose a challenge, but in this form it provides very little information; in essence it is a dead image. The challenge is to capture the symbolic representation of each character, so that it can be manipulated electronically. The current state-of-the-art systems can scan pages of text and translate the image into ASCII form at 98-99% accuracy. There are roughly 7.5 billion pages of text in the Library of Congress, which represents a sizable portion of world knowledge [50]. At a sustained rate of 10 seconds per page, an optimistic number, it would take 1000 such scanners 2.4

years to capture all of this information in ASCII form. Capturing the world's text would be a massive project, but it is still only a part of the data collection problem. Capturing three-dimensional representations of objects for a virtual museum collection would be an even bigger challenge.

9.3.2 Visualization

Scientific visualization involves the application of display technology, computers, and interaction, in order to gain a better understanding of a set of salient features. It relies on knowledge about human perception in order to make the best choices for displaying and collecting information. For example, in working with audio we may identify totally different concepts or salient features. One would probably decide to use hearing to judge the artistic quality of a musical piece, but one would alternatively prefer a graphical image when analyzing audio signals as frequency components.

The GROPE III system [36] is one of many efforts that use computer-based models and interactive displays to gain some understanding of how nature works at the molecular level. A closer look at the molecular modeling application will help underscore why human computer interfaces are so important. The earliest research that used computer displays to understand and model molecular structures was conducted at MIT in the mid-1960s [54]. Since then, computers have become powerful enough to run simulations of the bonding forces at interactive speeds for some simple models. A typical application for this technology is in the design of drugs, and the simulations help evaluate how well a drug molecule bonds to a particular site. Ideally, the computer models and algorithms could evaluate a drug's bonding potential on their own. However, the computer algorithms are inherently slow at this, because there are a large number of less-than-optimal solutions.

At UNC, researchers found that it was more efficient to put a human into the loop and let their recognition skills detect the best possible docking solution. Initial experiments used the ANL-3 six-degree-of-freedom manipulator and a large rear projection display shown in Figure 9.7. In this case, visual feedback was enhanced to highlight bonding forces, seen as bright lines that pop out from various locations. In later experiments, force feedback was used to augment visual feedback. One control study [55] indicated that six-dimensional molecular docking was 1.75 times faster with force-torque feedback

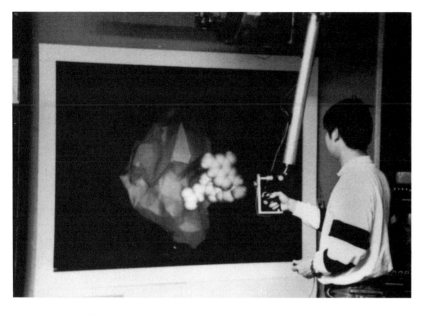

Figure 9.7
Visual and haptic feedback used to conduct molecular docking experiments at UNC. (Courtesy of M. Ouh-Young, AT&T Bell Laboratories.)

than without.[5] This is a perfect example of interactive visualization through parallel sensory modalities. A simulation of molecular bonding is the conceptual domain, and this process is translated into a force and torque representation that the user can feel and a visual representation that they can also see.

Some significant observations arise from this work and other types of visualization. There are many applications that benefit from having a human in the loop as the recognition and decision-making component. In scientific visualization, there are often no known solutions for an algorithm to target; instead the processes are used to help explore empirical data. The ultimate goal is understanding, through whatever means, and any modality or combination of senses should be exploited. Scientific visualization involves more than creating informative images.

Another established use for scientific visualization is medical imaging. The CAT scanner (computerized axial tomography) is a well-established tool for diagnosis and treatment planning. The raw output of these devices consists of a series of axial slices that cumulatively form a three-dimensional density field. This density field is primarily

---

[5]Think-time was 35% of the total docking time, and was subtracted out to yield the 1.75 improvement factor. The study also showed that drug trajectory path-lengths were 41% shorter with force-torque.

used to reveal structures inside a body. There is a correspondence between various densities and tissue types. Sometimes it is necessary to observe different layers of a density field. The layers correspond to transitions between different types of tissues. Here the extraction and enhancement of features is crucial. To view these layers as surfaces requires the use of algorithms to detect the layers and to map them into a surface representation [56,57].

There have already been reports of operations in which surgeons have used computer graphics to plan a procedure. In some cases, graphic displays were used as references during the operation. There is a possibility that this type of technology can be carried even further, to assist surgeons during surgery by providing information in ways that improve performance. The role of virtual reality and scientific visualization would be to do more than simply throw more information at the surgical team. The technology should also provide enhancement of the important features, and present the information in ways that do not create sensory overload. In the future, surgeons might use lightweight head-mounted displays to provide access to information during surgery. Surgeons commonly wear glasses that have magnification lenses attached. In a similar manner, the display might act like a pair of bifocals, where the surgeon can rapidly glance at vital statistics and surgical plans, see magnified views, or refer to x-ray images [58,59].

Computer-aided design is another field that will benefit from virtual reality and visualization technology. Today, there are many examples of simulation being used for computer-aided design and manufacturing. Computer simulation and analysis has become an essential tool for saving time and money in product design, because it eliminates the need for physical prototyping. Simulations allow designers to analyze different properties of a design under a variety of conditions, such as stress, temperature, aerodynamics, and so on. These properties can be used to compare the performance of different designs. Simulation can also be used to analyze the feasibility of a manufacturing process.

These forms of simulation are not obviously linked to virtual reality, because they are not interactive and may not require realistic depiction through three-dimensional displays. There is, however, a similarity in the underlying methods for representing and processing three-dimensional computer models. When computers are powerful enough to perform analysis at interactive speeds, these two fields will have a great deal in common.

Head-mounted displays can create the impression of being immersed in synthetic surroundings. This is extremely useful for examining architectural models. Viewing a model of a room with a standard CRT cannot reveal what it is like to actually be in the room. A model's proportions and their relationship to the scale of the human observer are more effectively perceived through a head-mounted display.

The need for simulated environments has largely been driven by applications to pilot training. Simulators can be demanding in terms of the technology. Flight simulators, for example, have traditionally been so expensive that the most practical application for them has been within military and commercial aviation.[6] Good simulators are expensive because they require realistic depiction of interactive artificial environments in real time. Real time typically refers to sustained update rates for IO devices. For the average simulator, the definition of real time would be the typical scanning rate of a CRT, 30-60 Hz. It is most common these days for people to use the term real time to describe interactive time (5-20 Hz), but ideally it should be used to describe the limits imposed by the display hardware. An important benchmark is the degree of latency. Latency refers to the time that elapses between an input event by the user and the output that depicts the consequences of that action. Every stage of processing adds to the total latency: input hardware, collecting input, updating the environment model, and finally displaying the updated environment. Human sensitivity to latency (especially when it fluctuates) is what makes closed-loop simulators difficult to implement.

Imaging is only one part of the technology that is required for effective flight simulation, and this will be true for other environment simulators. Motion cannot be perceived entirely by visual cues. This is why simulation cockpits are mounted on mechanical platforms that can be tilted to match the roll, pitch, and yaw that is visually presented. One major problem with simulators that attempt to simulate motion through space is their inability to generate the sensation of gravitational forces. The only practical solution for this problem has been the g suit, a body suit that uses inflatable bladders to imitate positive g forces.

Researchers at Stanford have proposed using computer models and virtual reality to create surgical simulators. A surgical simulator could help interns gain familiarity with some aspects of a procedure, with-

---

[6]As expensive as they are, they still save money. "Whereas it costs about $5000 per hour in fuel and maintenance to fly an F-16 jet fighter in training, it costs less than a tenth of that per hour to fly the F-16 simulator" [13].

out having to supply a cadaver or risk the health of a real patient. With the right sensor technology, every step of an operation could be recorded, capturing all events in a three-dimensional representation. The event recording would have to be at extremely high fidelity and would include: geometric structures, audio, forces, and so on. A simulator would replay the operation, allowing an intern to step into the surgeons shoes and experience the same events that the original surgeon did. If a sufficient level of detail existed in the computer models, the simulator could be used (like flight simulators are) to introduce a variety of hypothetical situations. The simulator could serve as a tool for instruction, as a means of evaluation, and as a tool for planning a surgical procedure.

9.3.3 Recreational

The entertainment industry is the most likely setting for many applications of virtual environments. This was how Huxley saw it, and present indications tend to support that view.

With head-mounted displays and tracking devices, the possibilities for creating realistic fantasy games are endless. Consider what a future dungeons and dragons game would be like. A force-sensing array in the floor tracks the contestant's steps along the cave floor. Three-dimensional foes attack with blows that look and sound harmful, when in fact they are completely harmless. With surround sound, monsters can be heard as they approach from behind. The synthetic dungeons could be filled with all sorts of magic, limited only by the imagination of the games designers.

This sounds great, but can it be done? Only flight simulators have the graphics capability to implement such vivid depiction of a fantasy game in real time. The costs of such systems will be a major factor in how such applications are implemented. The first implementations of these games may come as part of an arcade, where people rent the facility for a period of time. Eventually the cost may come down enough so that people would be able to buy their own virtual reality hardware like they buy video games today. After this stage is reached, the next will be a capability for personalized games that the end user helps define.

There are efforts underway in developing interactive versions of broadcast media. At the MIT media lab there is a group called the "Interactive Cinema Group." The group is involved in a variety of projects related to digital video and cinematography.[7] One project that is extremely relevant to virtual reality is being conducted by Michael Bove and others. He has demonstrated a method of gathering video images, along with range information, that uses depth from focus techniques. The ultimate goal of this work would be the creation of camera-like devices that can record three-dimensional movies. Movies or television in this form could be viewed from any vantage point.

Multiuser games would combine many of the techniques mentioned above with networking, enabling games to be played with human rather than computer-controlled opponents. For many people, multiuser games are inherently more interesting because the behavior of opponents is not as predictable. Games can be designed where people compete with each other, or they can be designed to promote collaboration.

Whatever level of realism can be simulated, these techniques could provide recreational services that go beyond games. A virtual reality glimpse at a vacation spot, home shopping, real estate, and so on, might provide useful services.

The acceptance of these types of service will undoubtedly depend on the nature of the medium and truth in advertising. It is not uncommon for the glossy brochure to give impressions of quality that are inflated. This might also be the case for virtual reality, but there may be some potential value in planning a vacation and shopping through such a medium. One could hopefully encounter these simulated vacations, products, or houses in *hypermedia* form. This might allow someone to quickly narrow down the choices – especially in cases like real estate, where an interactive tour might convey more accurate impressions.

9.4 CONCLUSIONS

The scope of virtual reality stretches across a wide variety of disciplines, both in basic enabling technologies and in its potential applications. This chapter has surveyed many of the devices and techniques as they have developed over time. Throughout the chapter, many of the research and development issues were discussed, but without

---

[7]This information was obtained from a booklet published by MIT in 1990, marking its 5-year anniversary.

any detail. The following is a summary of the relevant research and development issues.

Computer graphics technology is an essential element of virtual reality, particularly those that require high-resolution high-quality scenes in real time. Image synthesis has progressed rapidly, contributing new algorithms and computing architectures. The gaze-directed systems represent a solution that spans across many disciplines to achieve high-resolution synthetic images. A great deal of research in computer graphics has been concerned with creating more realistic scenes and realistic motions. For architectural simulations it is believed that *radiosity*, a method for simulating diffuse lighting conditions, is essential for properly judging computer-modeled interiors.

Simulation techniques will have to progress to the point were various properties (physical, geometric, etc.) can be modeled and used as part of an interactive media. Currently many simulations require supercomputers and cannot support interactive update rates. However, like computer memory chips, capacity and speed continue to go up as cost goes down. More efficient algorithms will also be required to make responsive systems.

Display devices continue to be a critical area of research for virtual reality. The devices that are used today reflect the desktop model of computing. Those that venture beyond this are still not in a form were they will be accepted in an office setting. So far, the Private Eye[8] display mentioned earlier is the only exception. Others are too heavy and too expensive and do not provide support for everyday tasks. Holographic displays are not the solution to every problem, but for many applications they may be the ultimate noninvasive form of display. All of the current approaches to synthetic holographic display have a long way to go before they can compete with conventional CRT or LCD displays.

IO devices for virtual reality pose many interesting challenges for researchers in a variety of disciplines. Effective tracking of rigid-body and non-rigid-body motion continues to be a challenging

---

[8] Private Eye is a registered trademark.

research and development topic. Both problems are most challenging when noninvasive techniques are required. In the use of audio, research has only scratched the surface, and there are many opportunities for new uses of audio feedback. An important form of audio will be synthetic speech. Speech recognition continues as a popular research topic, with the ultimate goal of 99.9% speaker-independent natural-language recognition being a long way off. Another research challenge involves those devices that provide input and feedback through the haptic senses. Within the next decade or two, these devices will challenge roboticists and mechanical engineers, as well as those who try to implement systems with them.

Integration of the various technologies will continue to pose other research and development challenges. In fact, integration is not a trivial part of the problem, because how these techniques are ultimately combined dictates many of the requirements of each piece of technology. Multimedia systems are still not commonplace in the office or even in the lab settings. Communications based on multimedia complicates matters even further, given the current state of the art in networks.

A variety of applications were mentioned with the intention of showing the potential impact that virtual reality may have. Examples were given in areas of communications, telerobotics, training, visualization, design, and entertainment. Some of these descriptions necessarily rely on extrapolation of the current technologies towards what might be possible in the near future. However, none of them relies on any major breakthrough, only improved performance levels and miniaturization of what is available today. It is almost certain that the first fruits of virtual reality lie just around the corner.

ACKNOWLEDGMENTS

Some of the ideas for applications have been influenced by and developed in collaboration with S.K Ganapathy. I also wish to thank him for his helpful suggestions regarding this chapter.

REFERENCES

[1] A. Huxley, *Brave New World*, Doubleday, New York (1932).

[2] I.E. Sutherland, The Ultimate Display, *Proceedings of the IFIP Congress*, pp. 506-508 (1965).

[3] I.E. Sutherland, A Head-Mounted Three Dimensional Display, in *Fall Joint Computer Conference*, pp. 757-764 (1968).

[4] L.G. Roberts, *The Lincoln Wand*, MIT Lincoln Laboratory Report (1966).

[5] A.M. Noll, *Man-Machine Tactile Communication*, Ph.D. Dissertation, Polytechnic Institute of Brooklyn, Brooklyn, New York (1971).

[6] J.J. Batter and F.P. Brooks Jr., GROPE-I: A Computer Display to the Sense of Feel, in *Proc. IFIP Congress*, pp. 759-763 (1972).

[7] D.L. Vickers, *Sorcerer's Apprentice: Head-mounted Display and Wand*, Ph.D. Dissertation, Department of Computer Science, University of Utah, Salt Lake City, UT (1974).

[8] R.A. Bolt, "Put-That-There": Voice and Gesture at the Graphics Interface, *ACM Comput. Graphics* **14**(3), 262-270 (1980).

[9] M. Krueger, *Artificial Reality*, Addison-Wesley, Reading MA (1983).

[10] M. Krueger, T. Gionfriddo, and K. Hinrichsen, VideoPlace: An Artificial Reality, in *CHI, Conference on Human Factors in Computing Sytems*, pp. 35-40 (1985).

[11] T.G. Zimmerman and J. Lanier, A Hand Gesture Interface Device, in *CHI+GI Conference Proceedings*, pp. 182-192 (1987).

[12] S.S. Fisher, M. McGreevy, J. Humphries, and W. Robinett, *Virtual Environment Display System*, ACM Workshop on Interactive 3D Graphics, Chapel Hill, NC (1986).

[13] R.N. Habor, R.N., Flight Simulation, *Sci. Am.* **255**(1), 96-103 (1986).

[14] M. Levoy and R. Whitaker, Case-Directed Volume Rendering, *Comput. Graphics: Symp. Interactive 3D Graphics* **24**(2), 217-223 (1990).

[15] M.A. Fischetti and C. Truxal, Simulating The Right Stuff, *IEEE Spectrum* **22**(3), 38-47 (1985).

[16] H.E. Ives, *An Experimental Apparatus for the Projection of Motion Pictures in Relief,* Bell Telephone System Monograph, **B-747** (1933).

[17] R.B. Collender, The Stereoptiplexer – competition for the hologram, *Information Display* **4**(6), 27 (1967).

[18] R.B. Collender, 3-D television without glasses: on standard bandwidth, *Proc. SPIE* 391 (1983).

[19] L.P. Dudley, Stereoscopy, in *Applied Optics,* Vol. 2, Rudolf Kingslake (ed.), Academic Press, New York (1965).

[20] J.A. Roese and A.S. Khalafalla, Stereoscopic viewing with PLZT ceramics, *Ferroelectronics* **10**, 47-51 (1976).

[21] A.C. Traub, Stereoscopic Display Using Rapid Varifocal Mirror Oscillations, *Appl. Optics* **6**(6), 1085-1087 (1967).

[22] J.N. Butters, *Holography and Its Technology,* Peter Peregrinus Ltd., London (1971).

[23] M.W. Levine and J.M. Shefner, *Fundamentals of Sensation and Perception,* Addison-Wesley, Reading, MA, pp. 397-400 (1981).

[24] L.J. Deutsch and A.M. Richards, *Elementary Hearing Science,* University Park Press, Baltimore MD, pp. 171-172 (1979).

[25] M.M. Blattner, D.A. Sumikawa, and R.M. Greenberg, Earcons and Icons: Their Structure and Common Design Principles, *Hum. Comput. Interaction* **B4**, 11-44 (1989).

[26] T.J. Doll, J.M. Gerth, W.R. Engelman, and D.J. Folds, *Development of Simulated Directional Audio for Cockpit Applications,* USAF report No. AAMRL-TR-86-014 (1986).

[27] E.M. Wenzel, F.L. Wightman, D.J. Kistler, and S.H. Foster, A Virtual Display System for Conveying Three-Dimensional Acoustic Information, *Proc. Hum. Factors Soc.* **32**, 86-90 (1988).

[28] D.F. Whightman and D.J. Kistler, Headphone Simulation of Free-Field Listening I: Stimulus Synthesis, *J. Acoust. Soc. Am.* **85**, 858-867 (1989).

[29] M.M. Blattner, R. Greenberg, and M. Kamegai, *Listening to Turbulence: An Example of Scientific Audiolization,* CHI Workshop

on Multimedia and Multimodal Interface Design, Seattle, WA (1990).

[30] S.R. Smith, D. Bergeron, and G.G. Grinstein, Stereophonic and Surface Sound Generation for Exploratory Data Analysis, in *CHI, Conference on Human Factors in Computing Sytems*, pp. 125-132 (1990).

[31] D.H. McBurney and V.B. Collings, *Introduction to Sensation/Perception*, Prentice-Hall, Englewood Cliffs,NJ (1977).

[32] M. Minsky, M. Ouh-young, O. Steele, F.P. Brooks, Jr., and M. Behensky, Feeling and Seeing: Issues in Force Display, *Comput. Graphics, Symp. Interactive 3D Graphics* **24**(2), 235-243 (1990).

[33] R.C. Goertz and R.C. Thompson, Electronically Controlled Manipulator,*Nucleonics* pp. 46-47 (1954).

[34] R.C. Goertz, et al., The ANL Model 3 Master Slave Manipulators - Its Design and Use in a Cave, in *Proceedings of the Ninth Conference in Hot Laboratories and Equipment*, Washington, DC (1961).

[35] P.J. Kilpatrick, *The Use of Kinesthetic Supplementation in an Interactive System*, Ph.D. dissertation, Department of Computer Science, University of North Carolina, Chapel Hill,NC (1976).

[36] F.P. Brooks, Project GROPE – Haptic Displays for Scientific Visualization, *Comput. Graphics* **24**(4), 177-185 (1990).

[37] H. Iwata, Artificial Reality with Force-feedback: Development of Desktop Virtual Space with Compact Master Manipulator, *Comput. Graphics* **24**(4), 165-170 (1990).

[38] F.H. Rabb, E.B. Blood, T.O. Steiner, and H.R. Jones, Magnetic Position and Orientation Tracking System, *IEEE Trans. Aerosp. Electromagn. Syst.* **AES-15**, 709-718 (1979).

[39] H.J. Woltring, New Possibilities for Human Motion Studies by Real-time Light Spot Position Measurement, *Biotelemetry* **1**, 132-146 (1974).

[40] R.P. Burton, *Real-Time Measurement of Multiple Three-Dimensional Positions*, Ph.D. Dissertation, Department of Computer Science, University of Utah, Salt Lake City, UT (1973).

[41] G. Bishop and H. Fuchs, *The Self-Tracker: A Smart Optical Sensor on Silicon*, Ph.D. Dissertation, University of North Carolina at Chapel Hill, Chapel Hill, NC (1984).

[42] J. Wang, V. Chi, and H. Fuchs, A Real-Time Optical 3D Tracker For Head-Mounted Display Systems, *Comput. Graphics: Symp. Interactive 3D Graphics* **24**(2), 205-215 (1990).

[43] J. Segen, GEST: An Integrated Approach to Learning in Computer Vision, in *Proceedings of the International Workshop on Multistrategy Learning – MSL-91* Harpers Ferry, WV (1991).

[44] B.A. Marcus and D.J. Sturman, Exotic Input Devices, in *Proceedings of NCGA'91*, pp. 293-299, Chicago (1991).

[45] J. Hong and X. Tan, Calibrating a VPL DataGlove for Teleoperating the Utah/MIT Hand, in *Proc. IEEE Int. Conf. Rob. Autom.* **3**, 1752-1757 (1989).

[46] L. Pao and T.H. Speeter, Transformation of Human Hand Positions for Robotic Hand Control, in *Proc. IEEE Int. Conf. Rob. Autom.* **3**, 1758-1763 (1989).

[47] D.M. Weimer and S.K. Ganapathy, A Synthetic Visual Environment with Hand Gesturing and Voice Input, in *CHI, Conference on Human Factors in Computing Sytems*, pp. 235-240 (1989).

[48] J. Kramer and L. Leifer, The TalkingGlove: An Expressive and Receptive Verbal Communication Aid for the Deaf, Deaf-Blind and Nonvocal, in *Proceedings of the Annual Conference on Computer Technology, Special Education, and Rehababilitation*, California State Univiversity, Northridge, CA, pp. 335-340 (1987).

[49] R.A. Bolt, Gaze-Orchestrated Dynamic Windows, *Comput. Graphics* **15**(3), 109-119 (1981).

[50] R.W. Lucky, *Silicon Dreams*, St. Martin's Press, New York (1989).

[51] T.B. Sheridan, Telerobotics, in *IFAC 10th Triennial World Congress*, Munich, FGR, pp. 67-81 (1987).

[52] M. Noyes and T.B. Sheridan, A Novel Predictor for Telemanipulation through a Time Delay, in *Proceedings of the Annual Conference on Manual Control*, NASA-Ames Research Center, Moffett Field, CA (1984).

[53] V. Bush, As We May Think, *Atlantic Monthly,*July, 101-108 (1945).

[54] C. Levinthal, Molecular Model-building by Computer, *Sci. Am.* **124**(6), 42-52 (1966).

[55] M. Ouh-Young, D.V. Beard, and F.P. Brooks, Jr., Force Display Performs Better than Visual Display in a Simple 6-D Docking Task, *Proc. IEEE Int. Conf. Rob. Autom.*, pp. 1462-1466 (1989).

[56] G. Herman and H.K. Liu, Three-Dimensional Display of Human Organs from Computer Tomograms, *Comput. Graphics Image Processing* **9**(1), 1-21 (1979).

[57] M. Levoy, Display of Surfaces from Volume Data, *IEEE Comput. Graphics Appl.* **8**(3), 29-37 (1988).

[58] W. Robinette, *Hip, Hype and Hope: The Three Faces of Virtual Worlds*, ACM SIGGRAPH, Panel Session (1990).

[59] S. Ditlea, Grand Illusion, *New York Magazine*, August 26-34 (1989).

Index

Numbers in *italics* denote a figure.

Acrotony:
  defined, 165
  phenomenon, 153
Active server, 221
Adaptive histogram equalization
    (AHE), 236, *237*
Additive noise, 71
Additive synthesis, 199
Adjacent ribbons, 50
Advection equation, 19–20
Aliasing, 195. *See also* Antialiasing
Alu sequences, 104
AMAP, 166
Amino-acid substitutions, 116
Anaglyphic stereo, 254
Antialiasing, 228
Apical meristems, 147, *148, 169*
Archimedian spirals, 79
Architectural analysis, 146
Architectural models, 147, *149,* 150
Architectural unit, 166
Arrow plots, 34
Artifacts, 206
Artificial reality, 247
Auxiblasts, 150
Avalanche masses, 135, 137
Axillary meristems, 147, *169*

Backface removal, 224
B-bisplines, 72, *74*
Beam splitters, *251*
Belousov-Zhabotinskii (BZ) reaction,
    46, 65, 71, *71,* 77, *78*
Binomial filters, 87
Binomials, growth process and, 161

Biphasic structure, 82
Block-iterative processes, 224
Brachyblasts, 152–153
Branching order, 150, 153
Branching processes, 146, *152,* 154,
    164–165
Bresenham algorithm, 69, 229
Bubbles, 31, *33*
Butterworth low-pass filter, 75

Cell colonies, 66
Cellular automaton, 46, *48*
Cellular differentiation, 66
Cessation law, 162
Chaotic advection, 16–19, 23
Chemical gradients, 66
Chemical waves, 66
Chrominance illusions, 207–208,
    207– 208
Circular line diagrams, 97
Circulation, 17–18
Closeness-of-fit parameter, 73
Clouds, 38
Codon bias:
    display of, 110
    policies, 35
Coherent structures, 28, *28*
Color palettes, *198*
Color representation, *13*
Combinatorial trees, 146
Compressed video data, 239
Computational fluid dynamics (CFD),
    8–9
Computer-generated holography
    (CGH), 37

Computer graphics, growth of, 2
Computing illusions, 208–209, *209*
Cones of confusion, 257
Contextual region, 236
Contour map, 68
Contour plot, 34
Contrast illusion, *207*
Cosmic jets, 28–30, *29*
Creeping flow, 21
Curl field, two-dimensional, *187*
Cyberspace, 245
Cylindrical slicing, 49–50, *49, 52–54,*
    57
Cytomegalovirus DNA, 106
Cytomegalovirus genome, 113

da Vinci, Leonardo, sketches from, *8*
DeBoor-Cox recurrence relation, 73
Deceptions, of display, 204, 206–210
De-interleaving, 224
Deoxyribonucleic acid (DNA):
    conduit, 93
    defined, 91
    line diagram, *96*
    loci, 97
    nucleotides, *100*
    sequences, 91–95, *112*
Diffuse reflections, 129
Diffusion constant, 19
Diffusion-limited (DLA) model, 140
Diffusion representation, *193*
Digital-signal processor, 217
Dilation, 126
Distribution functions, *158–160*
Divide-conquer-marry paradigm, 227
DNA, *see* Deoxyribonucleic acid
    (DNA)
Drainage basins, *18*
Droplet:
    coalescence, 124, 126, *128*
    configuration of, *133*
    deposition, *139*
    dimensionality of, 125
    formation of, *131*
    growth of, *127*
    sliding, 135
    on spider web, *141*

    stages of, *132*
    steady state, *138*
    three-dimensional, *134, 140*
    two-dimensional, off-lattice model,
        136

Earcons, 257
Elastic line, 166, 170–171
Euclidian substrates, 140
Euler equations, 16
Eulerian, 11
Excitable elements, 46, *47*
Excitable media, 45
Excited state, 48, *52*
Exons, 94
Eye trackers, 262–263

Fast Fourier transforms, 71, 75, *76.*
    See also 3D fast Fourier
    transform (3D FFT)
Feynmann diagrams, 93
Fick's law, 87
Flight simulation, 248, *250*
Flood-fill algorithm, 82
Flood-fill technique, 68, 77
Flow velocity, *33*
Fluid flows, 8
Fluorescence (LIF), 27–28, *27*
Foam, evolution of, 31, *32*
Forest tree, developmental sequence,
    151
Formal approach, virtual
    experimentation, 183
Fractal aggregates, 185, *187*
Fractal field, *204*
Fractal functions, *237*
Fractal geometry, *186, 237*
Fractal iso-surfaces, 34
Fractal landscapes, 185
Fractals, generally, 146
Frame buffer, 219
FRAMES system, 223
Functional domains, 94

Galaxies, collision of, *191*
G curves, 97–100
GenBank, 95

General Relativity theory, *192*
Geometric illusions, 206
Geometry Supercomputer Project, 2
Global mode of representation, 111
Graftals, 146
Graphics workstations, 3
Gravitational lens, 191
GRIPS (Graphic Raster Interactive Processing System), 67
Growth engine, 167, *171*
Growth unit (G.U.), 147

Hand trackers, 262
Haptic displays, 257–260
H curves:
    applications, 103–111
    bundle, 116–117, *117*
    DNA nucleotide, *105*
    DNA sequences, *112*
    G curves and, 98, 100–102
    low resolution, *106–107*
    short, *101*
    stereoscopic diagram of, *109*
HDTV (High-definition television), 210, 265
Head-mounted displays, 246, 249–252, *251*
Heat transfer, 135
Hidden lines, 69
Holograms, 37, 255
HRTFs (Head-related transfer functions), 257
Hue, 35
HYLAS, 102
Hypercube, 99
Hypermedia, 265
Hypertext, 265
Hypertexture, 238, *238*

Image convolutions, 224
Image processing, *236*
Importance sampling, 240
Inflorescence, 154
Intensity profile, 79, 81–82, *82–83*
Interleaving, 215, 224
Internode:
    defined, 147

growth and, 154
Introns, 94
Inviscid flow, 19
Isometrics, G curve and, 99
Iso-surfaces, 34

Julia set, 237, *237*

Kinematic viscosity, 19
Knots-adding procedure, 73
Knotted ribbon, 50, *51–53*

Lactate dehydrogenase (LDH), 85, *85*
Lagrangian, 11
Lambda bacteriophage, 103, *103*
Lambert-Beer's law, 70
Landscape, rendering of, *173*
Lenticular stereo, 253, *254*
LIFE, 46
Light intensity, *81*, 85–86
Lincoln Wand, 246
Line rasterizing, 229–230, *230*
Line tree, growth pattern, *172*
Local averaging, 49
Local nucleotide composition, 94
Loudness, 256
Low-pass spectral filtering, 68, *76*
Luminance, 200. *See also* Luminosity
Luminance illusion, 207
Luminosity, 68

Mach bands, 206–207, *207*
Mandelbrot set, *189, 206*
Material lines, 20
Meristem activity, *155*, 161
Mesoblasts, 152
Mesotony, 165
Metabolites, 94
Micro-mechanical structures, 260
MIMD (Multiple Instructions Multiple Data), 214
Modon, *205*
Molecular docking experiments, feedback used, *267*
Monochromatic palettes, 200
Monopodial branching, 147
Monte Carlo method, 167

Morlet wavelet, *203*
Morphological descriptions, 146
Moving average, 71, *72, 80*
Multimedia system, 239
Multiplanar technique, 254–255

NADH (Nicotinamide adenine
    dinucleotide), 85, *85*
National Science Foundation, 3
Navier-Stokes equation, 19
Nerve axons, 45
Neurospora crossa, H curve, *104*
Nonrandom codon usage, 110
Non-von Neumann bottleneck, 227
Nullclines, 47
Numerical approach:
    divergence display, *188*
    virtual experimentation, 183

Oligonucleotides, 110
Optical illusions, 195, 206
Order-by-order simulation, *168*
Order number, 150
Orthotropic, 147

Paracladial systems, 146
Parallel architecture, 214
Parallel computing, 240–241
Parallel growth generation, *170*
Passive server, 221
Pathlines, 12–*13*, 35
Performance analysis, 226–228
Phong's shading model, 27
Photorealistic rendering, 225
Physiological age, 150
Pipelining, 215
Pipe nodes, 215–216, *216*
Pixel funnel, 216
Pixel Machine, 214, 216–217, *216*
Pixel nodes:
    block diagram of, *217*
    communication between, 224
    computation in, 223–224
    system architecture and, 216
Pixel noise, 68
Plagiotropic, 147
Planar sectioning, 49–50, *50*

Planar slicing, 49–50, *54–58*
Poincare section, *17*
Point of origin coding, 19
Poisson process, 157
Polychromatic palettes, 200–201
Polygonal graphics, 241
Polygonal rendering, 228
Polygon rasterizing, 230–231, *231*
Polymer strings, *190*
Prefix order simulation, *168*
Preformed part, 154
Primary operations, chain of, 196–197,
    197
Procaryotic cell, 94, 97
Program overlays, 226
Projection operations, 197, 199
Projective geometry, 8
Proleptic branching, 164
Proprioception, 257
Protons, *190*
Prunus avium:
    branching system of, *152*
    growth of, *175*
    tree simulation, *175*
Pseudo-colors, 68
Purine, 101
Pyrimidine, 101

Queuing theory, 156

Radiosity, 272
Raster operations, 229, *229*
Rayleigh-Taylor instability, 25
Ray-sphere intersection, 224
Ray tracing:
    generally, 232–233
    illustration of, *233–234*
    models of, 129
Receptive state, 45, 48
Reductions, 199
Reference axis, 167
Reflectance calculation, 129
Refractory state, 45, 48
Regulatory processes, 65–66
Reiteration, 146, 150
Resolution, 36–37
Rest probability, 161

Restriction endonucleases, 94
Reynolds number, 16, 34
Ribbons, 50, *51*
Ribosomal RNA (rRNA), 115
Riemann surfaces, 93
RISC processor, 218

Saccade, 263
Saturation, 35
Scaled arrows, 77
Scaling form, 126
Scaling symmetry, 124, 126
Scientific visualization:
    defined, 2
    generally, 266–270
Scroll waves, 45, 52
Secondary operations:
    chain of, 197
    three-dimensional representation,
        203
Self-pruning, 153
Semi-random automaton, 46
Sensation level, 256
Sequence annotations, 113
Shading correction, 70
Signal-to-noise ratio, 68
SIMD, 214–215
Simultaneous contrast, 206
Smoke, 36, 38
SMOOPY, 72–73, 75
Smoothing algorithms, 68
Soap film, 30–31, *30*
Sobel and Laplace filters, 71
Solids modeling, 25, *25*, 37
Solignac distribution, 160
Spectral filtering, 75
Spectrophotometer, two-dimensional,
    66–67
Specular reflections, 129
Sphereflake execution, plot of, *234*
Stereo images, 252–255, *253*
Stereoptiplexer, 253
Stereoscopic visualization, 99–100
Stokes flow, 16, 21, *21*
Streaklines, 14, 35
Streamlines, 14
Sylleptic branching, 164

Synthesis, 195
Synthetic turbulence, 23

Tactile display, 259–260
Teleoperation, 247
Teleoperator, development of, 258
Telepresence, 264
Telerobotics, 264, *264*
Temporal factor, picture synthesis, 195
Textual annotations, 113
3D fast Fourier transform (3D FFT),
    240
3-D graphics, 3
Threshholding, 235
Timbre, 256
Timelines, 14, *15*, 35
Tornado, 20
Toroidal image extension, 75
Torus network, 219
Transfer RNA (tRNA), 115
Transformations, 199
Translucency accumulation, 235
Transparent ellipsoids, 146
Turbulent flow, 14
2-D graphics, 3
Two-dimensional splines, 68, 71

Ultrasonic tracking, 260
Unknotted ribbon, 50, *51–52*

Value, 35
Vector normalization (shading), 224
Velhust dynamics, *188*
Velocity field, 10
Virtual display lists:
    stabilized platform-deployment
        station, *232*
    tea room, *232*
Virtual experimentation, 183
Virtual reality, 210, 245, 257, 265
Virtual texture maps, museum room,
    233
Viscous flow, 21
Visual clues, 202
*Visualization in Scientific Computing
    (ViSC)*, 2
Volume rendering, 235–236, *235*

von Karman vortex street, 12
Vortex dynamics, 19–20
Vortex pairing, 28
Vortex rings:
  colliding, 23, *24–25*
  fluorescence and, 27
  leapfrogging, 22–23, *22*
Vortices, *26*, 28
Vorticity:
  equation, 19

  magnitude, 19
  secondary operations and, 202

Waterspout, 20
Water waves, 38
Wire-frame images, 37

Z-buffer access, 223
Zollner illusion, 206–207, *206*